Ex dono Dr. RD Hutchins
olim commensalis
2000.

Submillimetre wave astronomy

Edited by
J.E. BECKMAN and J.P. PHILLIPS
Department of Physics, Queen Mary College, London

CAMBRIDGE UNIVERSITY PRESS
Cambridge
London   New York   New Rochelle
Melbourne   Sydney

Published by the Press Syndicate of the University of Cambridge
The Pitt Building, Trumpington Street, Cambridge CB2 1RP
32 East 57th Street, New York, NY 10022, USA
296 Beaconsfield Parade, Middle Park, Melbourne 3206, Australia

First published 1982

Printed in Great Britain at the University Press, Cambridge

Library of Congress catalogue card number: 82-4487

British Library Cataloguing in Publication Data

Submillimetre wave astronomy
1. Radioastronomy ⸺ Congresses
2. Submillimetre waves ⸺ Congresses
I. Beckman, J.E. II. Phillips, J.P.
522'.682 QB474

ISBN 0 521 24733 0

CONTENTS

PREFACE

The submillimetre wavelength range has been aptly called the last
frontier in observational astronomy.   During the 1960s and 1970s
radioastronomers were purposefully extending their ability to observe to
higher and higher frequencies, while infrared astronomers were pushing to
longer and longer wavelengths with equal diligence.   The interval
between them was whittled down to the submillimetre range, roughly between
1 mm and 100 µm wavelength, which posed special problems of technique,
difficulty of observation and astrophysical interest.

Infrared observers have traditionally used photoconductors
or bolometers to detect faint infrared sources.   Cooled to liquid helium
temperatures, these devices are used either broad-banded, or with optical
style spectrometers.   Spectrophotometry performed this way is best
suited to observing the dust component of the interstellar medium,
although atomic and ionic studies of importance have been achieved
particularly in the near infrared.   Radio techniques, on the other hand,
are characteristically those of superheterodyne detection, most effective
over the narrowest spectral bands.   These are used to best effect in
observing individual spectral lines, and hence as probes of the interstellar gas.   In the submillimetre range these techniques overlap,
giving the range a potentially key role in the analysis of the relatively
cool interstellar clouds where stars are forming.   It has been all the
more frustrating that the water vapour rotational band in the earth's
atmosphere cuts out a great proportion of the incoming radiation between
100 µm and 1000 µm, so that observations have been made through a series
of rather murky windows scattered within the range.

Several developments have now come together to produce an
explosion of submillimetre possibility.   New telescopes at high
altitude, especially those on Mauna Kea, Hawaii, at last offer sites

where eighty per cent of the ground-level water vapour path is below the
observer.   Balloons and aircraft give the submillimetre observer as free
a run, albeit for restricted periods of time, as the ground-based
observer in the visible or the radio regions.   Within a few months, the
IRAS satellite should be providing the first unbiased survey of point
sources at the short wavelength (100 μm) end of the range.   Within a few
years large ground-based telescopes (30 m Franco-German, 15 m British-
Dutch, 10 m U.S.) will allow observations at 1 mm wavelength with
angular resolution better than that now confined to the 100 μm range.

  The astrophysicists who attended the Submillimetre Wave
Astronomy Conference held at Queen Mary College, London, in September
1981 (with support from the Royal Society, and under the auspices of
the Royal Astronomical Society) reflected the position of the subject
at the nexus of discipline and technique.   There were infrared and
radioastronomers with an admixture of theoretical chemists specializing
in astrophysical problems.   While the subject is moving away from the
phase where advances in technique dominate astrophysical results,
specialists in the newest techniques formed an important contingent.

  A key topic current in submillimetre astronomy is the analysis
of giant molecular clouds on a galaxy-wide scale.   The first section of
the book reflects this strong interest, and shows where major advances
are being made.   Using the CO molecule as the prime tracer ,clouds of
molecular hydrogen at temperatures below 50 K, sheltered from direct
starlight by their internal dust, can be mapped throughout the galaxy.
As CO allows radial velocity to be measured directly, the classical
methods developed by Oort enable these cloud complexes to be positioned
within the galaxy.   We know that these clouds form the matrix out of
which new star clusters form.   We also know that their lifetimes are
short, that they dissociate within 10^7 years or so, and that new clouds
must form in order to replenish the stock from which the spiral arms
themselves can continually regenerate.   An attack on all of these
phases is now being mounted by observers and theorists.   Improvements in
our understanding of the formation of the clouds themselves, how they are
distributed in the galactic plane, and how the conditions within them
become ripe for star formation are all represented here.

  In the second section we are looking with spectroscopic methods
at individual clouds and cloud groups.   Some of these, the Orion and
the M17 clouds for example, are favourites with observers, since they

contain many different molecular species, are large, bright and
therefore relatively easy to observe.   In this section the coming
together of infrared and radio techniques is seen at its most striking
with the former reaching up to 120 μm wavelength as observed from aircraft, and the latter down to 430 μm as seen from the ground.   Here also
we see work on the outer atmospheres of expanding stars, and detections
of external galaxies.   These two submillimetre fields are at their
inception, and will undoubtedly increase rapidly in importance in the
near future, as the big telescopes bring improved angular resolution.

        Section three describes molecule formation, and shows the
increasingly fruitful collaboration between those coming from backgrounds
in chemistry and in astrophysics.   Laboratory spectroscopists have been
able to point the direction to observers by predicting frequencies and
intensities of new species, and observers have responded by detecting tens
of molecules and radicals, together with their isotopically substituted
variants.   An important programme is to relate the observed abundances
of these species in the interstellar medium to those predicted from
theoretical models for the formation of their constituent elements.
The intervening chemical processes (reactions in the gase phase or on the
surfaces of dust grains) make it difficult to relate these observations
directly to stellar evolution theory or, for deuterium and helium to
conditions within the primaeval cosmic fireball, but the prizes in
improving our fundamental understanding in these two areas are great.

        We round off the volume with a short section on instrumentation
which picks out salient development points.   Most important here is the
fostering of new techniques in high frequency radioastronomy.   New local
oscillators, based on principles related to that of the maser, new
cryogenic mixers based on the Josephson and related quartum effects, and
new spectrometers ("backends") based on acousto-optic techniques are in
the course of development.   When used on the new large telescopes these
will enable the next generation of observers to extend to external
galaxies the kinds of studies represented, for our own galaxy, in this
book.

        Queen Mary College has been an active centre for submillimetre
astronomical research since the 1960s when the subject began.   It was
an appropriate venue for the conference, and provided an agreeable setting
for the presentation of a lively picture of an exciting field of research.

                                                          March/April 1982

SECTION I

Large-scale structure and radiative transfer
within interstellar clouds

THE FORMATION OF GIANT CLOUD COMPLEXES

B.G. Elmegreen
University of Sussex, England, and
Columbia University, New York, USA

Abstract. Various formation mechanisms for giant
cloud complexes are reviewed. We show as an example
how the giant molecular clouds and OB associations
in the Orion, Perseus and Sco-Cen regions could have
formed as condensations in Lindblad's expanding ring.
A new model for the ring is proposed which includes
these features in addition to several other prominent
cloud complexes in the solar neighborhood. We also
discuss the formation of star complexes and the triggering of cloud formation by spiral density waves.
A distinction between primary and secondary cloud
formation mechanisms is emphasized.

Giant molecular clouds form by a combination of processes that operate on very large scales in the interstellar
medium. These clouds are associated with known spiral arm
tracers, namely the OB associations (Lada et al. 1979), and
they appear to highlight the spiral arm structure better than
H I (Cohen et al. 1980). To the extent that the ratio of
molecular to atomic hydrogen increases in the spiral arms over
that in the interarm regions (Cohen et al. 1980), molecular
clouds must form in the arms, and most of them must get de-
stroyed before they reach the next arm. A theory of cloud
formation must explain how a spiral arm can trigger their
growth, and how the clouds evolve once they form.

There are two basic categories for the various mechanisms of cloud formation: primary mechanisms form first-
generation clouds out of a quiescent interstellar medium, and
secondary mechanisms form later-generation clouds after stimu-
lation from processes associated with the first generation of
clouds. These two mechanisms probably operate simultaneously
in all spiral galaxies, including our own, although the

relative importance of each mechanism may vary from galaxy to
galaxy, or with the phase of a spiral wave in any one galaxy.
As an example, consider a first generation of stars whose H II
regions, winds and supernovae exert a pressure on the cloud
that formed it, and on the surrounding interstellar medium.
Such pressures can disrupt a primary cloud and move it aside,
and they can sweep up a large shell or ring in the interstellar
medium.   Secondary cloud formation occurs when the first generation cloud recollapses, or when the swept-up shell collapses.
Observations indicate that the distances between primary and
secondary generations of star formation can be 100 to 300 pc
or more, and the time intervals between generations can be 20
to 100 million years.

          Evidence for secondary cloud formation is all around
us.   One of the best examples occurs in giant star complexes
(Efremov 1979).   A star complex is a composite of several adjacent and nearly contemporary star clusters, as defined by
their common halo of red supergiants and Cepheid variables.
Star complexes extend for several hundred parsecs along the
galactic plane, and they include stars with an age span of up
to 50 or 100 million years.   A good example of a star complex
is the aggregate of young clusters and dense molecular clouds
within 200 to 300 pc of the double star cluster h and χ Persei.
The youngest clusters in this region are IC 1805, IC 1848, and
IC 1795 (which are also the giant radio sources W5, W4, and W3,
respectively).   The older stars in h and χ Persei formed about
20 million years ago (Schild 1967), IC 1848 and IC 1805 began
forming stars about 5 million years ago (Stothers 1972), and
IC 1795 was provoked into forming stars less than 1 million
years ago, by the expansion of the H II region around IC 1805
(Lada et al. 1978).   The entire complex shows indirect evidence
for (1) a first generation of dense clouds that must have been
present in the past to produce h and χ Persei, (2) the subsequent removal of these clouds from the vicinity of h and χ
Persei, and (3) a second generation of dense clouds now observed
to be forming IC 1848, IC 1805, and IC 1795.   The clustering
of these star formation sites is probably not coincidental:
Efremov (1979) catalogued about 35 such regions.

The sequential star formation process (Elmegreen & Lada 1977) that formed IC 1795 differs from the mechanism of cloud regeneration discussed here: sequential star formation occurs in single clouds over small regions (1 to 50 pc) and in short time intervals (1 to 10 million years); i.e., it occurs during the disruption of a single cloud. In contrast, cloud regeneration occurs over a much larger region and in a longer time interval after the previously disrupted cloud and any associated shells of matter recollapse into new clouds.

More direct and spectacular evidence for cloud regeneration on a large scale comes from H I and CO observations of the solar neighborhood. The sun appears to be surrounded by a large, expanding ring of gas and dust that was discovered first as a source of H I emission by Lindblad (1967), and later as a source of CO emission by Cohen et al. (1980). Dame and Thaddeus recently mapped the nearby side of this ring (as reported by Professor Thaddeus at this conference), and they found that it coincides with the Great Dark Rift in the Ophiuchus-to-Cygnus region of the Milky Way. The Great Rift is therefore a molecular and atomic cloud, with a physical size similar to that of other giant molecular clouds (100 pc), and an age too young or density too low to be forming stars at the present time. The Great Rift appears to have been swept up by the same pressures that formed the Lindblad ring.

The Lindblad ring may be responsible for forming more than the Great Rift, however. The OB associations in Orion (Ori OB 1), Perseus (Per OB 2), and the Oph-Sco-Cen region are located near the periphery of the Lindblad ring, and the average radial velocities of the Orion and Perseus molecular clouds are similar to the expansion velocities of the Lindblad ring at the same positions. Figure 1 shows the distribution of all the prominent clouds in the solar neighborhood, with the cloud sizes drawn approximately to scale. The perspective is that of an observer outside the galactic plane. To make this figure, we have included: (1) light solid lines: the large-scale distribution of dust within $\pm$ 50 pc of the plane, plotted as contours of equal gas density with contour values of 1.2 cm^{-3}, 2.5 cm^{-3},

5 cm^{-3}, and 10 cm^{-3} (from Lucke 1978, Figure 8, with a conversion of dust to gas from Jenkins and Savage 1974). (2) Cross-
hatched circles: all of the OB associations within 500 pc of
the sun are drawn with circle sizes equal to their largest projected diameters (from Blaauw 1964). (3) Light dashed ellipse:
a simple model for the Lindblad ring (from Lindblad et al.1973).
(4) Heavy oval: the model for the Lindblad ring proposed here.
(5) Circle-dot: the position of the sun. Various features that
we believe to be associated with the Lindblad ring are labeled
with Roman numerals and listed in Table 1.

> Figure 1: Various components of the interstellar
> medium near the solar neighborhood are superposed.
> The light contours outline dust clouds (Lucke 1978),
> the cross-hatched circles are OB associations (Blaauw
> 1964), the dashed ellipse is the model for an expand-
> ing H I ring proposed by Lindblad et al. (1973), and
> the two heavy lines represent the model for the ring
> proposed here. The features indicated by Roman num-
> erals are described in Table 1.

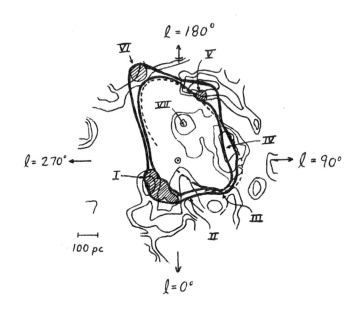

The origin of the Lindblad ring is unknown, but its expansion age is around 6×10^7 yrs (Lindblad et al. 1973). The ring expands so slowly (i.e., 4 km s^{-1} plus a component from galactic differential rotation) that the stars or star cluster originally responsible for the energy input to the ring could have already moved far outside of the ring. At a modest velocity of 10 km s^{-1}, the pressurizing cluster could now be more than 2 ring radii away from the ring's center. In order to make such a ring, the hypothetical cluster would have to deposit a large amount of energy ($\gtrsim 10^{51}$ ergs) into the interstellar medium in a time short ($\lesssim 10^7$ yrs) compared to the cluster crossing time over a ring radius; the corresponding kinematic power input would have been 10^{51} ergs/10^7 yrs$\gtrsim 10^3$ L$_\odot$. Perhaps a burst of OB star formation in a cluster that was once located in the center of the ring delivered the required kinetic energy to the local interstellar gas.

Let us examine Figure 1 and Table 1 closely. The model for the Lindblad ring proposed here is very similar to

TABLE 1: Prominent Young Features Associated with Lindblad's Ring

Region	ℓ	D	Description
Ring Components:			
I	289.6	155	IC2602
	292-312	160	Lower Centaurus-Crux OB association
	312-341	170	Upper Centaurus-Lupus OB association
	341-2	170	Upper Centaurus-Sco OB2 association
II	390-15	120-200	Ophiuchus molecular cloud and a large dust cloud
III	0-50	100-250	Great Dark Rift in Milky Way
IV	90-120	200-300	Large dust cloud
V	150-170	320-420	Per OB2 association, giant molecular cloud, and a large dust cloud
VI	199-210	460	Ori OB1 association, giant molecular cloud, and a dust cloud
Ring Interior:			
VII	150-185	110-220	Taurus molecular clouds and a small dust complex

the model by Lindblad et al. (1973) except in the direction of the galactic center. Aside from this direction, we have drawn the preferred model to pass through the large dust complexes at galactic longitude $\ell = 105°$ and distance from the sun D=250 pc (Region IV), at $\ell = 160°$ and D=350 pc (Region V, which is also the Per OB 2 association), and at $\ell = 210°$ and D=500 pc (Region VI; also the Ori OB 1 association). We have also drawn the ring to lie at the curving boundary of the dust complex between $\ell = 0°$ and $50°$ with D between 100 and 250 pc (Region III; identified here as the Great Rift in the Milky Way).

The portion of the ring drawn by Lindblad et al. (1973) in the direction of the galactic center is revised here to pass through the Oph-Sco-Cen region. We make this revision for several reasons: first of all, the H I data used by Lindblad et al. have a large gap between $\ell = 240°$ and $\ell = 300°$ and another gap between $\ell = 310°$ and $\ell = 330°$. Elsewhere the H I is well sampled. These gaps are represented in our figure by breaks in the light dashed ellipse. The gaps are suspicious because the CO data (Cohen et al. 1980) between $\ell = 10°$ and $\ell = 60°$ show the Lindblad feature at slightly higher velocity (8-12 km s^{-1}) than do the H I data (4 - 8 km s^{-1}) in this longitude interval. Thus the densest (i.e., CO-emitting) parts of the ring appear to be moving slightly faster than the low-density H I in the direction of the galactic center. Unfortunately there is no CO data between $\ell = 240°$ and $\ell = 330°$. Nevertheless, the CO velocity at $\ell = 10°$-$60°$ is the same as the mean radial velocity of the Sco-Cen association, which is 10 km s^{-1} (Petrie 1962). Such a large peculiar velocity for the nearby Sco-Cen association prompted Petrie (1962) to "interpret the motion as a general drift away from the sun corresponding, perhaps, to the motion of a cloud in which stars presumably originated". We agree with this interpretation; the dense gas that formed the Sco-Cen association apparently broke off from the H I ring as the ring slowed down, and in doing so, preserved a slightly larger velocity than the lower density H I gas around it. A second reason for revising Lindblad's model is that the nearest piece of the ring would be only 15 pc from the sun if the model were strictly correct. No evidence for any prominent H I or CO feature at this small distance has been found, but there is a large H I sheet covering $40°$ in the sky in this direction (Uranova 1959; Crutcher & Riegel 1974). The distance to this sheet can be estimated (in principle) from the presence or lack of optical absorption lines in the spectra of stars with known distances. Data from Olano and Pöppel (1981) are consistent with our model where the nearby portion of the ring passes through the giant dust complex that sur- rounds the Sco-Cen-Oph association. We also note that the ring model given by Lindblad et al. assumed for simplicity that the ring expanded freely, like a system of ballistic par- ticles. No consideration was given to the effect of pressure during the expansion, nor to any snowplow-like decelerations that would be expected. Nevertheless, the basic model by Lind- blad et al. (1973) appears to be quite good, needing only minor revisions in the direction of the galactic center.

The ring drawn in Figure 1 appears to have condensed into 3 separate OB associations: Perseus (Region V), Orion (Region VI) and Sco-Cen (Region I). The Great Rift in the Milky Way (Region III) and the prominent dust condensation at Region IV are along low column density portions of the ring, and they have not yet begun to form stars. The dust cloud near the center of the ring (Region VII) contains the Taurus dark cloud complex. These dark clouds have a low total mass, and they contain only young stars; they probably reached their high densities only recently (i.e., within 10^6 yrs).

The solar neighborhood illustrates well the processes of secondary cloud formation. These processes seem to have occurred as follows: (1) A first generation star cluster formed 60 to 70 million years ago near the center of the present Lindblad ring. This first cluster may have formed at the same time as Gould's Belt (which has a similar age -- Tsioumis and Fricke 1979) by a primary mechanism that operates on a very large scale (1kpc). (2) During a burst of massive star forma- tion, the first cluster disrupted most or all of its primordial cloud and deposited enough energy into the interstellar medium to create eventually a large cavity and expanding ring.

(3) 20 million years ago the ring began to collapse along its
periphery, first forming the Sco-Cen and Orion OB associations
(which are about 20 and 12 million years old respectively,
according to Blaauw 1964), and later forming the Per OB 2 association (which is 4 million years old according to Blaauw 1964).
We note that Stothers (1972) derives an age for each of these
associations of between 10 and 20 million years, so the ring
could have collapsed simultaneously into three major pieces.
These major sites of star formation are identified here as the
second generation. They are still near the remnants of the
expanding ring because the dense clouds and the ring always
had about the same velocities. (4) Today, star formation persists in the Orion molecular cloud, in the Ophiuchus cloud near
the Sco-Cen association, and in NGC 1333 and various other
clouds near Per OB 2. Each of these remnants of the current
generation of star formation has been kept active by sequential
star formation (Elmegreen & Lada 1977). Low mass star formation occurs in these OB associations (Gratton 1964), and also
in some of the smaller bits of cloud that are left over inside
the ring, like the Taurus clouds. (5) Finally, we note that
new systems of expanding shells are beginning to grow around
Orion (Reynolds & Ogden 1979; Cowie et al. 1979), Sco-Cen
(Weaver 1979; Heiles et al. 1980), and Perseus (Sancisi 1970;
Loren 1976), possibly setting up a third major generation of
star formation 20 to 50 million years from now.

  Secondary cloud formation mechanisms are dramatic
and, in some cases, obvious once the observer knows what to
look for: giant shells in the interstellar medium are common
(Heiles 1979), and star formation along the periphery of some
giant shells is easily observed (Goudis & Meaburn 1978). In
contrast, the mechanism of primary cloud formation remains a
mystery. Observations of this mechanism may be difficult for
several reasons: primary clouds may not be forming near the
sun because the sun is located between the major spiral arms
(see Simonson 1976), or the mechanism may operate too slowly or
on too large a scale to be observed unambiguously. Nevertheless,
theoretical considerations give us some idea of what could trigger a first generation of clouds when a spiral density wave

passes by. One possibility is that the interstellar medium
is unstable to forming large clouds because of a combination
of the Parker (1966) instability and the Jeans instability.
The Parker instability is a magnetic Rayleigh-Taylor instability
in the galactic gas layer, and the Jeans instability is driven
by the self-gravitational forces of the gas. Previous work on
the Parker instability by Mouschovias et al. (1974), and on
the Jeans instability by Elmegreen (1979), point to the likely
role that these processes play in forming first-generation
clouds.

An instability that is driven by the combination of
magnetic and self-gravitational forces, designated the Parker-
Jeans instability, has been analyzed in two recent papers
(Elmegreen 1982a,b);   the principal results of these two papers
will be summarized here. One important aspect of the Parker-
Jeans instability is that the driving forces depend sensitively
on the ambient gas density. Clouds tend to form near the galactic midplane and in regions of the galaxy where the ambient
density is highest. At low density, only magnetic and cosmic
ray pressures drive the instability, but at moderate and high
densities, the self-gravitational forces contribute. The density where self-gravity becomes important is about 1 cm^{-3};
such a critical density is, in fact, equal to the average density in the local interstellar medium (as determined by the
rate of extinction along the galactic plane;   see Spitzer 1978).
The average interstellar medium lies midway between the regimes
of magnetic force dominance and gravitational force dominance.

When the density of the interstellar medium lies below about 1 cm^{-3}, as it may between the major spiral arms, the
Parker-Jeans instability will grow slowly, on a time scale of
20 to 50 million years. (Coriolis forces will resist but not
stop the growth of this instability because the growth rate is
nearly equal to the epicyclic frequency in the galaxy). The
clouds that form in such a low density medium are relatively
low in mass (10^{4} to 10^{5} $M_{\odot}$), and, since this pure Parker instability is driven entirely by pressure forces in the interstellar medium, the clouds will form at subsonic velocities,
and their final equilibrium densities will be only 3 to 5 times

greater than the ambient density (see the non-gravitating equilibrium solutions by Mouschovias 1974). Thus the low-density interarm gas may form low density clouds slowly. These clouds may be shocked into forming stars by a spiral density wave, or they may not form stars at all. In contrast, the large densities expected (Roberts 1969) and observed (D. Elmegreen 1980) in spiral wave shocks (i.e., 3 to 5 cm^{-3}) can cause the Parker-Jeans instability to grow rapidly, in only 12 million years or less. The resulting cloud masses will be large, $10^6 M_\odot$, and the clouds will be significantly self-gravitating from the start of their formation. Such primary clouds will be ripe for condensation and fragmentation into smaller and denser molecular clouds as the gas cools.

This transition from the pure Parker instability at low, interarm densities, to the Parker-Jeans or pure Jeans instabilities at high spiral-arm densities could be the nec-essary stimulus given to primary cloud formation by a spiral density wave. Cloud regeneration then maintains star formation in an arm once the primary clouds form.

References

Blaauw, A. (1964). The O Associations in the Solar Neighborhood. Ann.Rev.Astron.Ap., 2, 213.
Cohen, R.S., Cong, H., Dame, T., Thaddeus, P. (1980). Molecular Clouds and Galactic Spiral Structure. Ap.J., 239,L53.
Cowie, L.L., Songaila, A. & York, D.G. (1979). Orion's Cloak: A Rapidly Expanding Shell of Gas Centered on the Orion OB 1 Association. Ap.J., 230, 469.
Crutcher, R.M., & Riegel, K.W. (1974). Optical Interstellar Line Studies of a Nearby Cold Cloud. Ap.J., 188, 481.
Efremov, Yu.N. (1979). On the Nature of Star Complexes. Sov. Ast.Lett., 5, 1.
Elmegreen, B.G. (1979). Gravitational Collapse in Dust Lanes and the Appearance of Spiral Structure in Galaxies. Ap.J., 231, 372.
Elmegreen, B.G. (1982a). The Parker Instability in a Self-Gravitating Gas Layer. Ap.J., 253, 634
Elmegreen, B.G. (1982b). Cloud Formation by the Parker-Jeans Instability. Ap.J., 253, 655
Elmegreen, B.G., & Lada, C.J. (1977). Sequential Formation of Subgroups in OB Associations. Ap.J., 214, 725.
Elmegreen, D.M. (1980). An Optical Analysis of Dust Complexes in Spiral Galaxies. Ap.J.Suppl., 43, 37.
Goudis, C., & Meaburn, J. (1978). Four Supergiant Shells in the Large Magellanic Clouds. Astron.Ap., 68, 189.

Gratton, L. (1974). Associations and Very Young Clusters. In
            Star Evolution. Proceedings of International School
            of Physics, Enrico Fermi Course No. 28, ed. L. Grat-
            ton, p. 243, New York: Academic Press.
Heiles, C. (1979). H I Shells and Supershells. Ap.J., 229, 533.
Heiles, C. Chu, Y.H., Reynolds, R.J., Yegingil, I. & Troland,
            T.H. (1980). A New Look at the North Polar Spur.
            Ap.J., 242, 533.
Jenkins, E.B. & Savage, B. (1974). Ultraviolet Astronomy from
            the Orbiting Astronomical Observatory. XIV. An Ex-
            tension of the Survey of Lyman α Absorption from
            Interstellar Hydrogen. Ap.J., 187, 243.
Lada, C.J., Elmegreen, B.G., Thaddeus, P. & Cong, H.I. (1978).
            Molecular Clouds in the Vicinity of W3, W4, W5.
            Ap.J.Lett., 226, L39.
Lada, C.J., Blitz, L. and Elmegreen, B.G. (1979). Star Form-
            ation in OB Associations. In Protostars and Planets,
            ed. T. Gehrels, p. 341. Tucson: University of Arizona
            Press.
Lindblad, P.O. (1967). 21-cm Observations in the Region of the
            Galactic Anticenter. BAN 19, 34.
Lindblad, P.O., Grope, K., Sandqvist, Aa. & Schober, J. (1973).
            On the Kinematics of a Local Component of the Inter-
            stellar Gas Possibly Related to Gould's Belt. Astron.
            Ap., 24, 309.
Loren, R.B. (1976). Colliding Clouds and Star Formation in
            NGC 1333. Ap.J. 209, 466.
Lucke, P.B. (1978). The Distribution of Color Excesses and
            Interstellar Reddening Material in the Solar Neigh-
            borhood. Astron.Ap., 64, 367.
Mouschovias, T.C. (1974). Static Equilibria of the Interstellar
            Gas in the Presence of Magnetic and Gravitational
            Fields: Large Scale Condensations. Ap.J., 192, 37.
Mouschovias, T.C., Shu, F.H. & Woodward, P.R. (1974). The
            Formation of Interstellar Cloud Complexes, OB Assoc-
            iations and Giant H II Regions. Astron.Ap., 33, 73.
Olano, C.A. & Pöppel, W.G.C. (1981). Neutral Hydrogen Emission
            Features in Scorpius and Ophiuchus and the Origin of
            Sco OB 2. Astron.Ap., 95, 316.
Parker, E.N. (1966). The Dynamical State of the Interstellar
            Gas and Field. Ap.J., 145, 811.
Petrie, R.M. (1962). The Scorpio-Centaurus System and the
            Absolute Magnitudes of the B stars. Mon.Not.Roy.Ast.
            Soc., 123, 501.
Reynolds, R.J & Ogden, P.M. (1979). Optical Evidence for a
            Very Large Expanding Shell Associated with the I
            Orion OB Association, Barnard's Loop, and the High
            Galactic Latitude Hα Filaments in Eridanus. Ap.J.
            229, 942.
Roberts, W.W. (1969). Large Scale Shock Formation in Spiral
            Galaxies and its Implications on Star Formation.
            Ap.J., 158, 123.
Sancisi, R. (1970). Motion of Neutral Hydrogen in Connection
            with the Association II Per. Astron.Ap., 4, 387.

Schild, R. (1967). Ages and Structures of Stars in the h and χ
        Persei Association. Ap.J., <u>148</u>, 449.
Simonson, S.C. III (1976). A Density Wave Map of the Galactic
        Spiral Structure. Astron.Ap., <u>46</u>, 261.
Spitzer, L. Jr. (1978). Physical Processes in the Interstellar
        Medium. New York: Interscience.
Stothers, R. (1972). Fundamental Data for Massive Stars Com-
        pared with Theoretical Models. Ap.J., <u>175</u>, 431.
Tsioumis, A. & Fricke, W. (1979). A Contribution to the Kine-
        matics of Gould's Belt. Astron.Ap., <u>75</u>, 1.
Uranova, T.A. (1959). Light Absorption in a Region in the Rift
        of the Milky Way. Sov.Ast. <u>3</u>, 718.
Weaver, H. (1979). Large Supernova Remnants as Common Features
        of the Disk. <u>In</u> Large Scale Characteristics of the
        Galaxy, ed. W.B. Burton, p. 295. Dordrecht: Reidel.

COSMIC RAYS AND GIANT MOLECULAR CLOUDS

A.W. Wolfendale,
Physics Department, University of Durham, Durham DH1 3LE

Abstract   Gamma rays have been observed coming from the direction of known giant molecular clouds, the quanta having almost certainly been produced by the interaction of cosmic rays with the atomic nuclei in the clouds.   Arguments are presented claiming that the cosmic ray intensity is known in those clouds within about 2 kpc of the sun and it is thus possible to take the γ-ray fluxes and evaluate the masses of the clouds.   It is found that the recipe given by Blitz (1980) relating CO data to column densities of molecular hydrogen gives a good description of the results. Application of the conversion factor to the CO surveys indicates that $\sim$ 30% of the gas in the Galaxy is in molecular form.

1   INTRODUCTION

The question of the mass of gas in molecular clouds in the Galaxy is an interesting one, bearing as it does on a variety of Astronomical parameters.   Derivation of the masses of individual clouds from the determination of the flux of millimetre radiation (e.g. the $J = 1 \rightarrow J = 0$ transition of ^{12}CO at 2.6 mm) is known to be very uncertain (see, for example, the recent review by Lequeux, 1981).   Here we adopt a quite different technique of mass estimation using the measured flux of γ-rays from the directions of specific clouds, the γ-rays being presumed to come from the interactions of ambient cosmic rays with the atomic nuclei in the clouds.

In previous work this type of analysis has been used in reverse form, that is, the masses of individual clouds have been assumed and the measured γ-ray fluxes have been used to estimate the cosmic ray intensity in the clouds (see, for example, Issa and Wolfendale, 1981; to be referred to as I).   It is still hoped that this method will, in the long term, prove valuable – i.e. to use molecular clouds as probes of the CR intensity – when purely astronomical techniques can be used to give precise cloud masses.   At present, however, we can choose a restricted region of space where the cosmic ray intensity is known with a degree of certainty and estimate the cloud masses as described in the first paragraph.

In earlier work (e.g. Dodds et al., 1975; Issa et al., 1981) we have used atomic hydrogen data (for which the mass conversion is believed to be accurate), together with γ-ray measurements to derive the conclusion that there is only a modest gradient of cosmic ray intensity in the Galaxy. For example, for the precursors of γ-rays above 100 MeV, a 50% mixture of electrons and protons (the electrons having median energy 300 MeV and the protons 3 GeV) we claim that the intensity changes by a factor 2 over a distance from the sun of 2-3 kpc, the factor being an enhancement towards the G.C. and a reduction towards the A.C. Thus, by restricting attention to the region within 2 kpc from the sun and taking clouds distributed roughly uniformly in longitude, we might expect the mean intensity to differ from that at the earth by less than about 20%.

At this stage it is relevant to point out that a detailed examination of radio synchrotron radiation by Phillipps et al. (1981) suggests a more complex dependence of cosmic ray intensity (electrons in this case) on position in the Galaxy. If equipartition between cosmic ray energy density and magnetic field is assumed (and e/p is constant) then the electron intensity can be determined as a function of position; application to the molecular clouds considered here indicates that the mean electron intensity is within 20% of the value at the earth (by 'at the earth' we mean after correction for solar and geomagnetic modulation).

It remains to consider the likelihood of there being enhancements in cosmic ray intensity inside the clouds considered – significant enhancements would, of course, invalidate the arguments. It is likely that some at least of the flux of low energy ($\overset{\sim}{<}$ 1 GeV) cosmic rays arises from objects associated with molecular clouds (OB associations, SNR ...) but the question is the extent to which the cosmic rays remain in the vicinity of the clouds before escaping. Examination of this topic is still in its infancy but some evidence comes from studies of the Orion cloud. Analysis of COS B data by Caraveo et al. (1980) and of SAS 2 data by Wolfendale (1980) indicated a γ-ray flux of the expected order of magnitude for this object and a more sophisticated analysis has been made recently by Issa and Wolfendale (1981b). These workers considered the ratio of the γ-ray fluxes (COS B) from the two components, Orion A and B, and compared this with what would have been expected for the alternatives: the clouds being inert, and there being enhanced cosmic ray intensities in A and B such as would have arisen if the various likely particle sources in A and B generated cosmic rays which were completely trapped in the clouds. Figure 1 shows the results. It is clear that there is no evidence favouring efficient trapping; the ratio of γ-fluxes is just what would be expected if the clouds were inert or, an indistinguishable situation, there were cosmic ray sources in the clouds which did not contribute significantly to the ambient intensity. A virtue of this analysis is that the ratio of the masses used is independent of the conversion from CO data to mass.

The case for extrapolating the results for Orion to other nearby clouds is that, to quote I, Orion has 'many flare stars, T-Tauri common; rich in O-associations'. Few of the other nearby clouds appear to have larger numbers of potential cosmic ray sources and thus we expect most of the clouds considered to be effectively inert. (In contrast, the Carina nebula at 2.7 kpc, considered in I and which appears to have a cosmic ray intensity enhanced by $\sim$ 20, is described as 'an unusual object; star in nebula is one of the most luminous in the Galaxy').

In what follows, a brief description is given of the local clouds selected (a more detailed description is contained in I) and their adopted trial masses are given, these masses following from the procedure described by Blitz (1980). This is followed by an examination of the γ-ray data and comparisons are made between the observed γ-ray fluxes and those predicted using the trial masses.

2 THE SELECTED LOCAL MOLECULAR CLOUDS

There is as yet no homogeneous catalogue of molecular clouds, the masses of which have been determined by a single technique. However Blitz (1980) has summarized data on 6 of the 10 clouds selected for the present work (three were measured by Blitz and the remainder by colleagues Cong, Chin, Kutner and Tucker). These satisfy the condition of being within 2 kpc of the sun and are in fact classed by Humphreys (1978) as being in the 'local' spiral arm. They are also roughly uniformly distributed in longitude, as required. Table 1 (from reference 1) gives the details. Some discussion is necessary of the method

Figure 1

Ratio of cosmic ray intensity in the components A and B of
the Orion cloud (for details see Issa and Wolfendale, 1981).
'γ' denotes the ratio derived from the measured γ-ray
fluxes and the ratio of the masses.

The arrows on the right indicate the ratios expected if cosmic
rays are produced and trapped in the clouds, their intensity
being proportional to the number of stars etc. It is apparent
that the observed ratio is inconsistent with all the
predictions (except for ΣL (optical)), strongly suggesting that
the cloud is inert or at least that particles produced in
sources within the components of the cloud escape rapidly and
do not enhance the cosmic ray intensity therein.

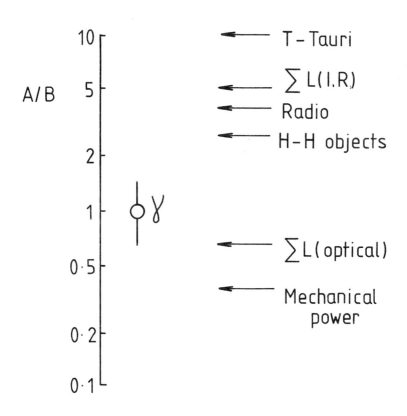

adopted by Blitz in order to determine the cloud mass – in the first instance, the LTE mass (LTE ≡ local thermodynamic equilibrium, the CO being assumed to be in this condition).

The 'mass' of a cloud M_{tot} is defined as the sum of three components: $M_{tot} = M_{obs} + M_{halo} + M_{uns}$. The first term, M_{obs}, comes from the observation of ^{13}CO, the lines from which are rarely self absorbed because of the low density of the ^{13}C isotope. M_{halo} comes from the observation of ^{12}CO in regions where ^{13}CO is too faint and M_{uns} is an estimate of the mass in the region where it is below the sensitivity limit. Typically, the mass is divided 23%, 55% and 22% amongst the three components. There follow corrections for channel dilution and non-linearity (typically 35%) to give $(M_{tot})_{corr}$.

It is now necessary to make an allowance for breakdown of LTE in the central, dense regions of clouds. Following the contention of Dickman (1978) that, for small dark clouds, the column densities are underestimated by factors of 2-3.5, Blitz adopted a factor 3; however the clouds of concern here are mainly rather large and we have usually adopted a factor (1-2) x only, with a best estimate of 1.5 x, i.e. M(best estimate) = 1.5 $(M_{tot})_{corr}$. Insofar as we have neglected contributions from H_1 and ionized gas (except for Mon OB2 – see Table 1) and their contribution could be 10% – 30%, the residual correction for non-LTE is seen to be rather small ($\sim$ 20-40%).

The above procedure relates to the 6 clouds considered by Blitz; for the remaining 4 we have endeavoured to make a similar analysis from published data.

3   THE PREDICTED γ-RAY FLUXES

The expected γ-ray flux from a cloud of mass M at distance d is $\Gamma_{\gamma,exp} = (q/4\pi)(M/d^2)$ where q is the γ-ray emissivity – i.e. number of γ-rays above 100 MeV for which Issa et al. (1981) quote $(q/4\pi) = 2.2 \times 10^{-26} s^{-1} sr^{-1} atom^{-1}$ with a cosmic ray intensity equal to that at the earth. The predicted γ-ray fluxes can now be derived assuming that the cosmic ray intensity in the clouds is the same as that at the earth (and also that all the cloud mass is accessible to cosmic rays – a reasonable assumption in view of the comparatively high energies involved). Table 1 gives the values.

Table 1   Data for the molecular clouds selected in the Analysis. F is the ratio of γ-ray flux observed to that expected for the adopted cloud mass.

Object	Distance (pc)	LTE mass of H_2 ($10^5 M_\odot$)	Adopted total mass ($10^5 M_\odot$)	γ-ray fluxes ($E_\gamma > 100$ MeV) $\times 10^{-6} cm^{-2} s^{-1}$ Predicted	Observed	F	Ref.
Mon OB1	800	1.0	1.5	0.6	0.3	0.5	1
Mon OB2	1600	1.0	4.0	0.4	<0.4	<1.0	1
C Ma OB1	1100	1.0	1.5	0.36	0.5	1.4	1
Orion OB1	450	2.0	3.0	4.2	1.4	0.36	1
Per OB2	350	0.4	0.6	1.4	<0.8	<0.6	1
Cygnus	1700	7.0	14.0	1.4	3.8	2.7	1
Gum	400	1.2	1.8	3.0	4.2	1.4	2,3
Cor. Aust.	150	0.07	0.1	1.3	0.4	0.3	4,5
Taurus	140	0.17	0.22	3.1	0.5	0.2	6,7
ρ-Oph.	160		0.07	0.7	1.1	1.6	8,9

References: 1, Blitz (1980); 2, Reynolds (1976); 3, Brandt et al. (1971); 4, Vbra (1977); 5, Rossano (1978); 6, Elias (1978); 7, Baud and Wouterloot (1980); 8, Bertiaud (1958); 9, Whittet (1974).

4 THE OBSERVED γ-RAY FLUXES

The data used come from the SAS 2 and COS B experiments
(Fichtel et al., 1978 and Thompson et al., 1976; and Wills et al.,
1980, Mayer-Hasselwander et al., 1980 and Hermsen, 1980). The derivation
of the γ-ray fluxes is difficult, in view of the poor angular resolution
of the instruments and the uneven backgrounds; however, such errors
as undoubtedly exist should be random and the mean of the final result
should be reasonably accurate. Details of the method of derivation of
the fluxes are given in I; here we give the results, in Table I.

5 COMPARISON OF THE OBSERVED AND PREDICTED γ-RAY FLUXES

Table I gives the ratios of the fluxes, F. In previous
work (I) these values were calculated for other clouds, further afield,
as well as those listed here, and were regarded as indicating an excess
of cosmic ray intensity compared with that at the earth where they were
significantly greater than unity. In the present case, for the reasons
given in §1, we take the cosmic ray intensity to be a constant and
attribute any deviations of F from unity to an error in the mass estimate.

The arithmetic mean of the F values is $\simeq 1.0$ and the median
is $\simeq 0.6$ so that the best estimate of F is $\simeq 0.8$. The spread of values
is roughly what we would expect from uncertainties in mass and γ-ray
flux.

6 CONCLUSIONS

The closeness of $<F>$ to unity implies that, with the assumptions made in the cosmic ray method, there is support for the Blitz
recipe for determining the masses of H_2 in molecular clouds within 2 kpc
of the sun and which have masses of order $10^5 M_\odot$, particularly when no
correction is applied for non LTE (the residual difference of 0.8 from
1.0 is then removed).

The extent to which this recipe can be applied to determine
the total mass of molecular gas in the Galaxy can now be considered.
Table 2 gives a summary of the situation with total mass as described
by Li and Wolfendale (1981). The values of the various parameters
adopted by different workers are indicated.

Insofar as the Blitz recipe uses $H_2/^{13}CO = 4 \times 10^5$ and
$T(^{12}CO)/T(^{13}CO) = 5.5$ there is seen to be some support for a total mass
of $\sim 8 \times 10^8 M_\odot$ (or $6 \times 10^8 M_\odot$ if the metallicity correction of Blitz and
Shu (1980) is taken) and taking the average of the Scoville and Solomon
(1975) and Gordon and Burton (1976) CO intensities ($\int Tdv$).

Table 2 Comparison of adopted quantities in the relation for column density of molecular hydrogen and the corresponding mass of H_2 in the Galaxy between R = 2 and 16 kpc

Reference	$\dfrac{H_2}{^{13}CO}$	$\dfrac{T(^{12}CO)}{T(^{13}CO)}$	Total galactic mass of H_2, $M_\odot$	Comments
Scoville and Solomon (1975) (SS)	2×10^6	5.5	3.2×10^9	
Gordon and Burton (1976) (GB)	6.67×10^5	3.0	2.0×10^9	
SS'	4×10^5	5.5	6.6×10^8	) prime denotes application) of metallicity correction) (cf Blitz and Shu 1980)
GB'	4×10^5	5.5	5.1×10^8	
SS"	4×10^5	5.5	8.7×10^8	) double prime denotes no) metallicity correction.
GB"	4×10^5	5.5	6.6×10^8	
Few (1979)	–	–	1.0×10^9	

(The column density of molecular hydrogen is given by

$$N(H_2) = \text{const.}\ \frac{H_2}{^{13}CO}\ \frac{T(^{13}CO)}{T(^{12}CO)}\ \int T(^{12}CO)\,dv\ )$$

In previous work (Issa et al., 1981) we considered the question of the local surface density of H_2 in some detail and concluded that $\sigma(H_2)/\sigma(H_1)$ is $\simeq$ 0.4, the analysis including the work of Few (1979) using formaldehyde and Lebrun and Strong (1981) using galaxy counts. Now, Gordon and Burton give data for H_1 and H_2 indicating an average value for the range R: 9.5 – 10.5 kpc of 0.8 so that the total Galactic mass of H_2 would fall from the GB value of 2 x $10^9 M_\odot$ to 1 x $10^9 M_\odot$, neglecting possible metallicity gradients.

Taking all available evidence a reasonable estimate of the mass of H_2 in the Galaxy is seen to be $\sim$ 8 x $10^8 M_\odot$, i.e. about $(^8/2.8)$ $\simeq$ 30% of the total mass of gas.

ACKNOWLEDGEMENTS

The author is grateful to the colleagues who work with him in the Gamma ray Astrophysics Group: Drs. M.R. Issa, Li Ti pei and P.A. Riley.

REFERENCES

Baud, B., and Wouterloot, J.G., 1980, Astr. Astrophys., 90, 297.

Bertiaud, F.C., 1958, Astrophys. J., 128, 533.

Blitz, L., and Shu, F.H., 1980, Astrophys. J., 238, 148.

Blitz, L., 1980, in "Giant Molecular Clouds in the Galaxy" (eds. Solomon, P.M., and Edmunds, M.G.), Pergamon Press, Oxford, 1., and Ph.D. thesis.

Brandt, J.C., et al., 1971, Astrophys. J. Lett., 163, L99.

Caraveo, P., et al., 1980, Astron. Astrophys. 91, L3.

Dickman, R.L., 1978, Astrophys. J. Suppl. 37, 407.

Dodds, D., Strong, A.W., and Wolfendale, A.W., 1975, Mon. Not. R. Astr. Soc., 171, 569.

Elias, J.H., 1978, Astrophys. J. 224, 857.

Few, R.W., 1979, Mon. Not. R. Astr. Soc. 187, 161.

Fichtel, C.E. et al., 1978, NASA Tech. Mem. 79656 (G.S.F.C., Greenbelt).

Gordon, M.A., and Burton, W.B., 1976, Astrophys. J., 208, 346.

Hermsen, W., 1980, Ph.D. thesis, Univ. Leiden.

Humphreys, R.A., 1978, Ap. J. Suppl., 38, 309.

Issa, M.R., and Wolfendale, A.W., 1981a, Nature, 292, 430.

Issa, M.R., Riley, P.A., Strong, A.W., and Wolfendale, A.W., 1981, J. Phys. G., 7, 973.

Issa, M.R., Strong, A.W. and Wolfendale, A.W., 1981, J. Phys. G., 7, 565.

Lebrun, F., and Strong, A.W., 1981 (in press).

Lequeux, J., 1981, Comments on Astrophysics, 9, 117.

Li Ti pei and Wolfendale, A.W., 1981, Astron. Astrophys. (in press).

Mayer-Hasselwander, H.A., et al., 1980, Ann. N.Y. Acad. Sci. 336, 211.

Phillipps, S., Kearsey, S., Osborne, J.L., Haslam, C.G.T., and Stoffel, H., 1981, Astron. Astrophys., 98, 286.

Reynolds, R.J., 1976, Astrophys. J., 203, 151.

Rossano, G.S., 1978, Astr. J., 83, 234.

Scoville, N.Z., and Solomon, P.M. 1975, Astrophys. J. Lett., 199, L105.

Thompson, D.J., et al., 1976, NASA Tech. Mem. X-662-76-198 (G.S.F.C., Greenbelt).

Vrba, F.J., 1977, Astr. J., 82, 198.

Whittet, D.C.B., 1974, Mon. Not. R. Astr. Soc., 168, 371.

Wills, R.D., et al., 1980, 'Non-solar gamma rays' (COSPAR, Ed. Cowsik, R., and Wills, R.D.) Pergamon, Oxford, 43.

Wolfendale, A.W., 1980, Origin of Cosmic Rays, IUPAP/IAU Symp. No. 84, (Eds. G. Setti, G. Spada and A.W. Wolfendale), Reidel, Dordrecht. 309.

FAR INFRARED LARGE SCALE MAP OF THE ORION REGION

E. Caux (1)
(1) CESR/CNRS-UPS - 9, avenue Colonel Roche - B.P. 4346
    31029 TOULOUSE CEDEX

G. Malinie (2), R. Gispert (3), J.L. Puget (2), C. Ryter (4),
N. Coron (3) and G. Serra (1)
(2) Institut d'Astrophysique CNRS- PARIS 75014
(3) LPSP-CNRS - VERRIERES LE BUISSON 91370
(4) CEA/CEN-SEP Ap. - SACLAY - GIF SUR YVETTE 91191

Abstract. A large area (7°x20°) around the Orion Nebula has
been surveyed in the far infrared range using a balloon borne
instrument with a 0.5° field of view. The instrument has been
especially designed to be sensitive to low brightness gradient
of extended sources. It works simultaneously in two wavelength
channels : 114-196 μm and 71-95 μm. The maps obtained in each
channel present two intense sources located respectively in
the direction of the Orion A/M42 Molecular Cloud/HII Region
Complex and the Orion B/NGC 2024 Complex.

The large beam and the scanning method allow an estimate of
the total power available in each Cloud-Complex.

The values of the ratio between the flux in the band 114-196μm
and the flux in the band 71-95 μm are different for the two
sources, showing differences in the dust temperature distribution.

INTRODUCTION

A lot of observations have been made in the Orion Region but
most of them focus on small areas, specially those located in the direction of HII Regions.

Particularly, in the far infrared range, observations by Harper
& Low 1971, Hoffmann et al. 1971, Emerson 1973, Fazio et al. 1974, have
shown that the Orion Molecular Cloud (M42, BN object, KL Nebula) is a
powerful FIR source.

These experiments usually have a spatial resolution smaller or
equal to 1 arcminute and are very sensitive to low fluxes but not to low
gradients of FIR brightness and such a low level IR emission over extended
regions cannot be accurately measured.

On the other hand, observations of CO line emission at 2.6 mm
have shown a very large extension of molecular cloud component over a few
degrees (Kutner et al. 1977).

In order to assess the total power emitted by all stars in the complex, the measure of the FIR brightness of the large molecular clouds extensions is necessary.

The present paper reports preliminary results about observations in the FIR range of a large area ($7°×20°$) in the Orion Region.

INSTRUMENTATION

Observations have been performed with AGLAE, the balloon-borne instrument previously described in Serra et al. 1978. This instrument has been especially designed to be sensitive to low FIR brightness gradients associated to extended sources (individual molecular clouds and Galactic disc).

The gradient sensitivity is $2.10^{-6} Wm^{-2} sr^{-1} degree^{-1}$ and point sources with fluxes greater than $10^{-6} Wm^{-2} sr^{-1}$ can be detected. The beamwidth is 0.5 degree.

The two Cassegrain telescopes of the instrument (15 cm diameter) focus the FIR beams onto the photometer. After the selection of two spectral bands by resthralen filters (TL1 for channel 1:114-196 μm and KI for channel 2:71-95 μm), the beams are focused with quartz lenses onto composite bolometers (Coron 1976).

The photometer is cooled with liquid helium (1.5 K) naturally pumped at the ceiling altitude. The telescope/photometer system oscillates around a vertical axis, scanning in azimuth at constant elevation ($h=20°$) with an amplitude of $10°$ and a period of 5s. This movement combined with the Earth rotation allows the observation of large areas of the sky.

The gondola is azimuth stabilized on the Earth magnetic field with an accuracy of a few arcminutes and the azimuth of the gondola can be changed from the ground by remote control.

This instrument has been launched four times : twice in FRANCE (1976 and 1977:Serra et al. 1978, 1979) 1 transmediterranean flight (1978 : Boissé et al. 1981, Gispert et al. 1982) and 1 flight in the south hemisphere (1980).

During this last flight (Alice Springs, AUSTRALIA 1980, November 12), the galactic ridge was observed from galactic longitude $260°$ up to $20°$ and several Molecular Clouds/H_{II} Region like ORION, ρ Ophiucus etc... were also mapped.

OBSERVATIONS IN THE ORION REGION

In this region two extended sources have been detected in the direction of the two following complexes :

Complex I   : Great Nebula of Orion, M 42, NGC 1932, Orion A, BN object, KL Nebula, OMC 1 and OMC 2.

Complex II   : NGC 2024, NGC 2023, ORIB, Horsehead Nebula.

The two sources have a larger angular size than the 0.5° beam and seem more extended in channel 1 (114-196 μm) than in channel 2 (71-95 μm).

Particularly, for complex I a large SE-NW extension (3°-4°) which is well spatially correlated with the ^{13}CO contours (Kutner et al. 1977) has been found.

The infrared maps in both channels are respectively shown on fig. 1 and 2.

The fluxes calibrations are obtained by comparison between the data obtained from the same sources observed in the last two flights (see Gispert et al. for details). Nearby molecular clouds complexes were also observed, for example in the Orion Region, the preliminary results of which are reported here.

In order to assess the total flux radiated by each complex one meets two difficulties.

First, the integrated flux of the low brightness extended wings is usually larger than the flux coming from the bright spots. The emission from these extended wings is very difficult to measure precisely.

Second, these extended wings often overlap and there is no obvious way to choose borders between individual sources. This problem is nevertheless harder in the galactic disc than for isolated complexes like those discussed here.

In Table 1 we give, for each complex, the fluxes coming from :

1 : the area centered on the brightest point of the source with an angular diameter equal to the beamwidth of the instrument,

2 : the area limited by the lower contour in the less sensitive channel (71-95 μm).

	COMPLEX 1			COMPLEX II		
	F1(114-156)	F2(71-95)	R=F2/F1	F1	F2	R=F2/F1
1	2.35	8.8	3.75	0.9	2.1	2.3
2	5.9	17.4	3.	1.3	2.7	2.1

Table 1 : FIR Fluxes in $10^{-9}Wm^{-2}$

A systematic difference between the two complexes can be noticed for the value of the ratio R. A dust temperature T of the source can be deduced from R, but it is very difficult to interpret such a value of T because the instrument measures an emission coming from a mixture of hot spots and a colder part of the cloud.

In order to give an indication and assuming a Q_{abs}-law in λ^{-2}, we can derive an average temperature of 60 K for the southern complex and 40 K for the northern one.

Gispert et al. have related the ratio R = F2/F1 with the IRE for each source.

Using flux density values at 5 GHz published by Goss and Shaver (1970) we can compute an IRE of 8 for the southern complex and 12 for the northern one.

In fact, the coldest parts of the cloud extend far away from the brightest parts of the FIR sources as can be seen in fig. 1 and as for the ^{13}CO contours. These large extensions contribute to a noticeable part of the power emitted in the FIR range. For example the total fluxes in channel 1 integrated over the lowest contour are $11.4 \ 10^{-9}$ for southern complex and $3.7 \ 10^{-9}Wm^{-2}$ for northern complex.

Assuming a distance of 450 pc for each cloud, and a ratio of 7.7 between fluxes in all FIR range and channel 1 (Puget, Serra 1982), we obtained $L_{FIR} = 5.6 \ 10^5$ L⊙ for all the southern complex and $L_{FIR} = 1.8 \ 10^5$L⊙ for all the northern one.

The total mass of gas in the clouds derived from the ^{13}CO low emission at 2.6 mm by Kutner et al. (1977) is 10^5 M⊙ for southern complex and $6 \ 10^4$ M⊙ for the northern one. We have only observed a part of the northern complex and if we integrate over the ^{13}CO contours in the area observed, we find approximately $4 \ 10^4$ M⊙. These values give an L_{FIR}/M ratio of ∿ 5 in solar units for both complexes. Such a value is not too

high compared to the average value expected in the solar vicinity
(1.5, Gispert et al. 1982).

· Thus, if we assume that the largest part of the dust is inside
the molecular clouds, the comparison between these two values : 1.5/5
seems to show that active star formation occurs in approximately one quarter of molecular clouds.

REFERENCES

Boisse P., Gispert R., Coron N., Wijnbergen J., Serra G., Ryter C.,
        Puget J.L. (1981) A & Ap. 94, 265.
Coron N. (1976) Infrared Physics 16, 411.
Emerson J.P., Jennings R.E., Moorwood F.M. (1973) Ap. J. 184, 401.
Fazio G.G., Kleinmann D.E., Noyes R.W., Wright E.L., Zeilik M. (1974)
        Ap. J. 192, L23.
Gispert R., Puget J.L., and Serra G. (1982) A & Ap., 106, 293.
Goss W.M., Shaver P.A. (1970) Aust. J. Phys. Astrophys. Suppl. n° 14,1-75.
Harper D.A.& Low F.J. (1971) Ap. J. 165, L9.
Hoffmann W.F., Frederick C.L. and Emery R.J. (1971) Ap. J. 170, L89.
Kutner M.L., Tucker K.D., Chin G. and Thaddeus P. (1977) Ap. J. 215,521.
Puget J.L. & Serra G. (1982) preprint
Serra G., Puget J.L., Ryter C. and Wijnbergen J.J. (1978) Ap. J. 222 , L21.
Serra G., Boissé P., Gispert R., Wijnbergen J., Ryter C. and Puget J.L.
        (1979) A & Ap. 76, 259.

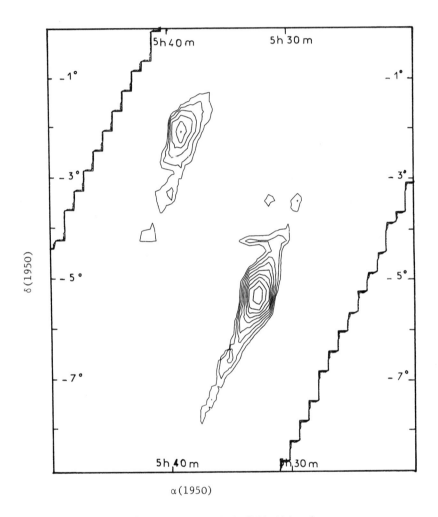

α(1950)

Fig. 1 : Channel 1 (114–196 μm)

First contour corresponds to
$4.7 \ 10^{-6} \ Wm^{-2} sr^{-1}$

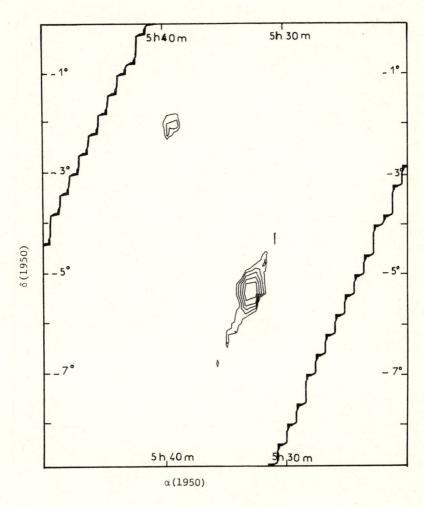

Fig. 2 : Channel 2 (71-95 μm)

First contour corresponds to
$2.3 \ 10^{-5} \mathrm{Wm}^{-2} \mathrm{sr}^{-1}$

COOL MOLECULAR CLOUDS

R.D. Davies
University of Manchester,
Nuffield Radio Astronomy Laboratories,
Jodrell Bank, Macclesfield, Cheshire, UK

Abstract The properties of the cool molecules in the diffuse clouds distributed throughout the interstellar medium are described. These clouds contain HI, H_2, CO, OH, H_2CO and other common molecules with temperatures of 10 to a few hundred Kelvins and with densities of 10 to 10^3 H atoms cm^{-3}. Optical absorption and UV studies indicate that dust is well mixed with the gas. Significant molecule formation only occurs in the local interstellar medium when the hydrogen column density is greater than 2×10^{19} atoms cm^{-2}. The giant molecular clouds represent a more massive but less numerous group of clouds.

Introduction

This paper reviews observational aspects of the molecule content of interstellar clouds. It will concentrate on the numerous clouds which characterize the general interstellar medium. These clouds have been studied at UV, optical (for their dust content) and radio wavelengths for their HI and molecular contents. They range in density from 10 to 10^3 H atoms cm^{-3} and in temperature from 10 to a few hundred degrees Kelvin. The discussion will make only passing reference to the dense cloud cores and to the giant molecular clouds (GMCs) in order to compare and contrast their properties. These clouds have often been labelled "diffuse" clouds to emphasize their broad distribution in the spiral arms and to contrast them with dense star-forming regions.

The environment of these clouds has also been considerably clarified in recent years. The clouds are imbedded in the spiral arms of the galaxy which in the vicinity of the Sun have a mean HI density of 1 cm^{-3}. About a quarter of this gas lies in a broadly distributed medium with a density of 0.24 cm^{-3} (Radhakrishnan, Murray, Lockhart & Whittle 1972). The low density HI clouds can be identified in emission by surveys at intermediate latitudes (e.g. van Woerden 1967, Heiles 1967). Typical HI densities are 14 cm^{-3}, cloud diameters are 7 pc and masses are 54 $M_\odot$;

approximately 11 such clouds are seen on a 1 kpc line of sight. Heiles found a larger spread in the cloud properties. The most common features were "cloudlets" with densities of 1-3 cm^{-3}, diameters of 3-6 pc and masses of 1-10 $M_\odot$.

A proportion of the emission line clouds are dense and cool and are also seen in absorption against background continuum sources. Surveys of HI absorption features provide information on the number density, optical depths and temperature of the clouds. These are molecule-bearing clouds as may be seen from the presence of identical absorption features in OH, H_2CO, CH, HCN and other widely distributed molecules.

UV observations have provided particularly useful quantitative information on the column densities of atomic and molecular hydrogen as well as of a number of atomic and molecular species along $\sim$100 lines of sight in the local spiral arm. These observations do not lead to the uncertainty in deriving H_2 column densities which is met in the radio case where the H_2 column density is inferred from the observation of CO or other molecular species and an assumed ratio of the molecule to H_2 density. It is interesting in all this work that there is a remarkably constant ratio of total hydrogen (atomic and molecular) density to dust density as measured by its optical absorption ranging from 0.1 to 10 magnitudes. Gas and dust seem to be well mixed in the interstellar medium in general and in the medium density clouds described here.

The existence or otherwise of interstellar molecules is the result of the competition between the formation and destruction mechanisms. Formation proceeds most quickly in dense regions whether it is the result of gas- phase or grain-surface reactions. Clearly destruction (dissociation) by UV or cosmic rays is greatest in the general interstellar medium and least within dense dust clouds. Various theoretical investigations indicate that molecule formation becomes significant when the obscuration is $\gtrsim$0.3 magnitudes.

Neutral hydrogen in diffuse clouds

Neutral hydrogen absorption spectra of galactic and extra-galactic radio sources reveal the presence of narrow absorption features which arise from diffuse HI clouds in the interstellar medium.

Measurements of these HI absorption spectra when combined with emission observations along adjacent lines of sight provide basic

information on the optical depth, spin temperature and velocity widths of
the diffuse clouds. Crovisier (1981) has analyzed recent high sensitivity
data which establishes a range in each of the parameters in contrast to
the concept of a "standard" cloud. Such absorption measurements can
sample conditions at large distances from the Sun in the galactic plane.

The mean velocity dispersion (σ) in the absorbing clouds is
~ 1.8 km s^{-1}. This dispersion includes both thermal and turbulent effects
which can be separated by using the direct determination of the spin
(thermal) temperature from a comparison of the emission and absorption
spectra. The mean Mach number in different samples of clouds is in the
range 1.1 to 1.8, indicating that turbulent motions are a significant
source of heating in diffuse clouds.

Crovisier's analysis of the spin temperature of clouds of
different optical depth is summarized in Table 1. The mean spin temperature falls with increasing optical depth to a mean value of 60 $\pm$ 4 K for
optical depths between 3 and 10. Table 1 includes an indication of the
HI line integral (N_H) and the density (n_H) within the clouds responsible
for the absorption features; these rough estimates are based on assumed
values of $\sigma = 1.8$ km s^{-1} and a cloud diameter of 10 pc. Values of N_H and
n_H for individual clouds are probably scattered by a factor of ~ 3 either
side of the mean values tabulated. Detailed studies of individual clouds
(e.g. Lockhart & Goss 1978) indicate that the parameters given in Table 1
are reasonable.

Table 1   Average parameters of HI clouds
(after Crovisier 1981)

Optical depth τ	Spin Temperature $\langle T_s \rangle$	HI line integral $\langle N_H \rangle$ (10^{19} atoms cm^{-2})	HI density $\langle n_H \rangle$ atoms cm^{-3}
$-2.0 < \log\tau < -1.0$	204	4	1.3
$-1.0 < \log\tau < -0.5$	126	15	4.8
$-0.5 < \log\tau < 0.0$	93	34	11
$0.0 < \log\tau < 0.5$	76	89	29
$0.5 < \log\tau < 1.0$	60	220	71

The frequency distribution of clouds as a function of optical depth is plotted in Fig. 1. This plot is adapted from the Arecibo and Nançay data given by Crovisier and shows the average number of clouds with optical depth (τ) greater than any given τ on a line of sight perpendicular to the galactic plane.   The ordinate is also given in terms of the average distance between such clouds on the assumption that the half-thickness of the galactic layer is 170 pc.   Fig. 1 gives the distribution of τ for the volume sampled by the survey, namely, the local spiral arm preferentially sampled at greater distances from the galactic plane. Lower latitude samples show an excess of strong features compared with the plot.   The absorption spectra of distant radio sources lying close to the galactic plane (e.g. Cas A, W49, W43, W51) show a frequency of $\tau > 1$ clouds greater by a factor of ~ 5 than in Fig. 1. A comprehensive survey of HI absorption spectra of low-latitude sources by Radhakrishnan & Goss (1972) shows 2.5 clouds per kpc averaged over arm and interarm regions; 80 percent of these clouds have $\tau > 0.5$. This suggests a smaller scale-height for the denser clouds and reflects the well-established observation that HII regions and the dense molecular clouds sampled in CO data have a

Figure 1   The frequency distribution of optical depth τ in HI absorbing clouds based on Arecibo and Nançay data (adapted from Crovisier 1981).   The right-hand scale gives the mean distance between clouds with the optical depth given

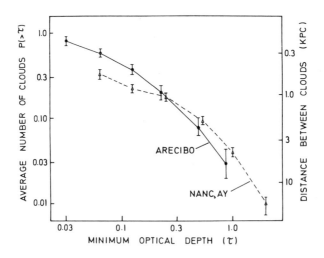

smaller scale height ($\sim$80 pc) than HI.

In addition to the "cool" HI clouds described above there is
a more uniformly distributed warm gas in the spiral arms. This HI comprises about half the mass of HI in the solar neighbourhood (Radhakrishnan
et al. 1972, Heiles 1967). A unit column of this gas centred on the Sun
and lying perpendicular to the galactic plane contains $\sim$1.5 x 10^{20} atoms
cm^{-2}. This gas has a range of temperature extending from a few hundred
to a few thousand Kelvins; it contributes substantially to the emission
spectra but only weakly to the absorption spectra.

Molecules in the local spiral arm - UV observations

Unambiguous estimates of the molecular hydrogen column density
along the line of sight to a star by evaluating $N(H_2) = \sum_{J} N(J)$ over the
J states of the H_2 UV absorption spectrum (Savage et al. 1977). Such
observations have been combined with HI column densities derived from Lα
absorption measurements to give the total hydrogen column density
$(N(H) = 2N(H_2) + N(HI))$ to 96 stars (Bohlin, Savage & Drake 1978). These
results show a close relationship between $N(H)$ and the colour excess E(B-V)
produced by dust on the line of sight to the stars. A best value of

$$N(H)/E(B-V) = 5.8 \times 10^{21} cm^{-2}\ mag^{-1} \qquad (1)$$

is representative of low and high density regions in the galactic plane
within 1000 pc of the Sun. A small upwards correction of $\sim$4 percent may
be required to take account of the ionized hydrogen in the local region.
The correlation between $N(H)$ and E(B-V) is shown in the lower left-hand
part of Fig. 2; E(B-V) is converted to A_V by adopting $A_V = 3.1E(B-V)$
(Savage & Mathis 1979). The UV results for this correlation supersede
those obtained using 21 cm observations in the direction of stars and
globular clusters (see for example Knapp 1974) where (a) some 21
cm emission may originate beyond the object, (b) the 21 cm emission may be
optically thick and (c) stray radiation effects in the 21 cm beam are
important for observations at high and intermediate galactic latitudes.
An important conclusion from the good correlation between $N(H)$ and E(B-V)
is that E(B-V) provides one of the best estimates of the column density of
the total hydrogen content. Deviations for a given line of sight are less
than a factor of 1.5 from the relationship given above (Bohlin et al. 1978).
Furthermore, the gas and dust appear to co-exist (Jenkins & Savage 1974).

40 percent of the stars observed in the UV have less than 1
percent of hydrogen in the molecular form. These lines of sight are
representative of regions where there are no dense clouds containing
molecules. The mean hydrogen density in this <u>intercloud</u> region is 0.16
cm^{-3}; the HI is distributed normal to the galactic plane with a gaussian
scale height of 350 pc (Bohlin et al. 1978). This mean density can be

<u>Figure 2</u>. A plot of the column density of interstellar gas
against extinction in the visual band, A$_V$. The column densities
are from UV measurements of HI and H$_2$, radio measurements of
^{13}CO and X-ray measurements. The relative scaling of the H
and CO data is discussed in the text.

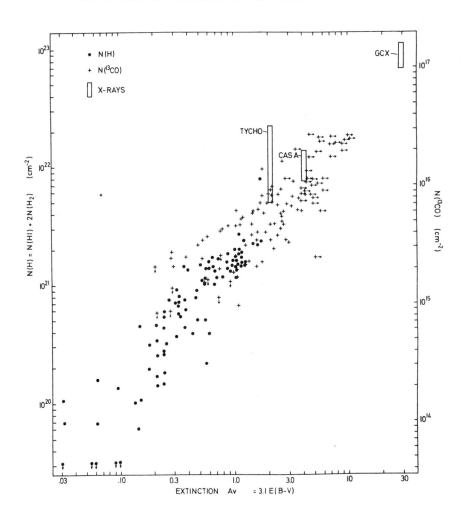

compared with 0.17 cm^{-3} and a gaussian scale height of 120 pc derived from
21 cm data (Baker & Burton 1975). Since both measurements are of the
line integral, the more accurate UV results show the presence of twice the
HI column density in the intercloud medium as determined at 21 cm wavelength.
The careful 21 cm absorption and emission results of Radhakrishnan et al.
give a value for the line integral in the distributed intercloud medium
in the solar neighbourhood which is close to the UV result.

If one uses the widely adopted value of colour excess of 0.61
mag kpc^{-1} for matter in the plane (Spitzer 1968) and the ratio N(H)/E(B-V)
given above, a value of the total gas density of 1.15 atoms cm^{-3} is
obtained for matter in the plane within 500 pc of the Sun. This total is
composed of 0.86 HI atoms cm^{-3} and 0.14 H$_2$ molecules cm^{-3}. These values
are probably representative of the state of hydrogen in the galactic
plane at a distance of 10 kpc from the galactic centre. The above values
contrast with those of Gordon & Burton (1976) who find n(HI) = n(H$_2$) =
0.4 cm^{-3} at this distance from the galactic centre; their results for
n(H$_2$) are based on CO observations and uncertain conversion factors.

UV observations of molecules and neutral atoms can be compared with similar data for H$_2$ in order to investigate the physical conditions within diffuse clouds (e.g. Federman et al. 1980, Federman 1981).
CO molecules have been detected in front of stars with A$_V$ $\leq$ 1.0 with
column densities up to $\sim 10^{15}$ cm^{-2}. When N(H$_2$) < 10^{19} cm^{-2} no CO is found
at a sensitivity of $\sim 10^{12}$ cm^{-2}. This total hydrogen column density appears
to define a threshold for the   formation of molecules and neutral atoms.
Above a value of $\sim 2 \times 10^{19}$ cm^{-2} the CO column density increases rapidly
with N(H$_2$) roughly as the second power. The column density of CO is more
tightly correlated with N(H$_2$) than with the sum of the atomic and molecular
column density. It would seem that the CO and H$_2$ are formed in interstellar clouds, while a substantial fraction of the HI lies in a distributed intercloud medium in which there are negligibly few molecules.

Molecular formation is a strong function of the local gas
density whereas the observations described above give only the line
integral of the molecules. Consequently a comparison between observation
and theory has to be based on model calculations with a realistic range
of physical parameters to describe the gas. Federman et al. (1981) have
used a model based on oxygen charge exchange chemistry using the cloud
parameters given in Table 2. Their high and low density models bracket

the observed values of N(CO) as a function of N(H$_2$)

More dense local clouds - the radio data

In clouds where the visual absorption approaches 1 magnitude, the informative UV observations of molecules described in the previous section become increasingly impractical. At higher values of absorption, radio observations of the 2.6 mm lines of ^{12}CO and ^{13}CO come into their own. Dickman (1978) has made a comprehensive study of CO in the solar neighbourhood in an investigation of 38 dark clouds with A$_V$ in the range 0.5 to 10 magnitudes. The ^{13}CO column densities were derived from ^{12}CO and ^{13}CO observations combined with an assumption of LTE. Calculations of a range of cloud models indicate that the LTE assumption leads to underestimates of N(^{13}CO) typically by a factor of 2. The observational data for N(^{13}CO) are plotted in the upper right-hand side of Fig. 2. The N(H) and N(^{13}CO) axes have been displaced by a factor of 0.7 x 10^6; a justification for this factor is as follows. Dickman finds the mean ^{13}CO column density to extinction ratio to be

$$N(^{13}CO)_{LTE}/A_V \ = \ (2.5 \pm 1.2) \times 10^{15} \ cm^{-2} \ mag^{-1} \qquad (2)$$

(this is strictly true for A$_V$ = 1.5 to 4 mag). Then using equation (1) and

$$A_V \ = \ 3.1 \ E(B-V) \qquad (3)$$

whence $N(H) = N(HI) + 2 \ N(H_2) = (0.72 \pm 0.36) \times 10^6 \times (N(^{13}CO)_{LTE}. \qquad (4)$

This calculation assumes that the ratio of N$_H$ to dust stays the same over

Table 2. Diffuse cloud parameters which are consistent with UV observations of neutral atoms and of CO

Parameter	Low density model	High density model
Average T (K)	55	30
Range of T (K)	30-75	25-35
Average n_H (cm^{-3})	150	2500
Range of n_H (cm^{-3})	70-300	2000-3000
CR ionization (s^{-1})	10^{-17}	10^{-17}
Carbon abundance	2.5 x 10^{-4}	2.5 x 10^{-4}
Oxygen abundance	3.5 x 10^{-4}	3.5 x 10^{-4}

the range $A_V \lesssim 1.5$, for which equation 1 is derived, to $A_V \sim 10$. Various arguments have been advanced that this assumption is valid and that the visual extinction is a good measure of the total hydrogen content of interstellar clouds covering the range of at least $0.1 < A_V < 10$ (Dickman 1978, Federman et al. 1980). Fig. 2 tends to confirm this suggestion. Further substantiation comes from X-ray observations of supernova remnants and the galactic centre source (e.g. Ryter, Cesarsky & Audouze 1975). Such observations show that a value of the ratio of $N(H)/E(B-V)$ given for equation 1 is probably appropriate out to $E(B-V) \approx 10$, the value for the galactic centre X-ray source. The X-ray values for the Tycho and Cassiopeia SNRs and for the galactic centre source are plotted in Fig.2.

CH is another widely distributed molecule in the galaxy which has been observed in diffuse clouds with a range of optical extinctions (Hjalmarson et al. 1977, Lang & Willson 1978). A comparison of optical and a range of radio ($\lambda \approx 9$ cm) transitions indicate that accurate estimates can be made of CH column densities. CH molecules would appear to be well distributed in diffuse clouds. The CH column density $N(CH)$ is well correlated with reddening and is given by

$$N(CH) = 6.3 \times 10^{13} \; E(B-V) \tag{5}$$

which on substitution in equation 1 gives

$$N(CH) = 1.1 \times 10^{-8} \; N(H) \tag{6}$$

This relation is well-established for $E(B-V) < 0.7$ and is a fair representation for $E(B-V)$ up to 2, although a lower coefficient will also fit the data in this higher range. Whether equation (6) continues to much higher values of extinction is unclear. The possible flattening of the relation indicated by the data at higher densities suggests that CO may be competing with the CH for the C atoms; further observations and modelling of the appropriate molecule-formation processes in clouds are needed to clarify the situation.

The molecular content of diffuse clouds in the local spiral arm has been investigated by many observers. OH, H_2CO, NH_3, HCO^+ and HCN are found in regions of optical obscuration where it may be assumed that H_2 is the most abundant molecule. In denser clouds with larger obscuration, UV observations are not possible and H_2 densities are estimated from the radio molecular observations combined with models which involve collisions between H_2 and at least some of the other molecules (e.g. Wootten, Snell & Evans 1980). Densities of up to 10^6 cm^{-3} are indicated in the most dense clouds. The volume density is a derived property and is less accurate than the column density which, apart from saturation corrections, is obtained directly from the observations.

Individual clouds and their relation to the giant molecular clouds

Among the smallest clouds identified are the Bok globules - compact dust clouds with high extinctions of up to 13 magnitudes. Martin & Barrett (1978) have made an extensive study of 12 globules in the radio lines of CO, CS, NH_3, H_2CO including several isotopic species. They found a strong positive correlation between the spatial extent of the molecules and the obscuring dust in each cloud. The kinetic temperature was constant across each cloud at $\sim$10 K while the derived densities were in the range $n(H_2) = (2.4 - 31.0)10^3 cm^{-3}$. The corresponding total gas masses of individual clouds were in the range 1.1 to 280 $M_\odot$. Martin & Barrett concluded that all the observed globules were gravitationally bound and were in the process of collapse from an interstellar HI cloud.

Both OH and H_2CO are easily detectable in dust clouds. Myers (1973) found a tendency for the OH emission to have a somewhat wider angular distribution and the H_2CO a narrower distribution than the dust. Heiles and Gordon (1975) found a good correlation between the antenna temperature of OH and H_2CO lines from place to place in the Taurus dust cloud. HI self-absorption followed the velocity pattern of the molecules suggesting that the atoms and molecules occupy the same overall volumes of the dust clouds, although the HI and molecules may not be co-located on the smallest scales. A constant ratio of $N(OH)/N(H_2CO) \approx 20$ is consistent with the Heiles & Gordon observations. Probably only 0.1 to 1.0 percent of the hydrogen is atomic in dense dust clouds (e.g. McCutcheon, Shuter & Booth 1978, Batrla, Wilson & Rahe 1981).

A number of investigations have used H_2CO as a major probe of the physical conditions in molecular clouds (see for example Wootten et al. (1980), Sandquist & Bernes 1980). Kinetic temperatures are found in the range 10 to 50K while densities are in the range $<10^4$ to 10^6 cm^{-3}. When dimensions and distances are available for these clouds, their masses can be estimated. These range from 1 $M_\odot$ for such clouds as L124 and L1529 to $\sim$10$^3 M_\odot$ for NGC1333, DR21(OH), S255 and OMC-2. The cloud associated with M17 has a mass of $\sim$3 x 10^4 $M_\odot$ and is probably a representative of the much less numerous class of giant molecular clouds (GMCs).

The existence of dense clouds beyond the solar neighbourhood
is demonstrated in several types of spectral line observation. Firstly,
absorption spectra of distant radio sources in the galactic plane show
the presence of such dense molecule bearing clouds. Fig. 3 shows the HI,
OH, CH and H_2CO spectra in the direction of the bright supernova remnant
Cas A. Two strong absorption features are seen in the Orion arm (V $\sim$ 0
km s^{-1}) and at least 4 in the Perseus arm (V = 40-50 km s^{-1}). These
clouds can be identified in the HI absorption spectrum also, indicating
that HI is either present in the clouds or is very closely associated
with the molecules in a surrounding envelope. This situation resembles
that in the dark dust clouds in the solar vicinity. A similarly close
correspondence between the OH, H_2CO and HI absorption spectra is found
(Fig. 4) in the direction of the non-thermal component (B) of W49 which

Figure 3   The HI, H_2CO and OH absorption spectra of Cassiopeia
A. The emission profile of CH shown inverted, is the result
of stimulated emission in the molecular clouds produced by the
radiation from Cassiopeia A. The main HI components are shown
dashed;   these are subject to more doppler broadening than the
molecular spectra.

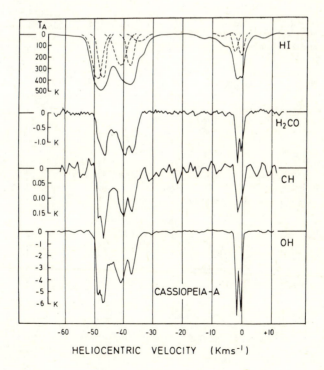

lies at a distance of ∿14 kpc.   In this case though there is more
variation in the relative intensities of the HI and molecular line depths.
These molecular absorbing clouds have typical velocity widths of a few
km s^{-1}.   By contrast, on some lines of sight through the Galaxy, a deep
absorbing feature is seen which has a velocity width of 5-10 km s^{-1}.   Such
broad deep features are seen in the absorption spectra of M17 and W51;
the central velocity of these features are close to those of the HII
regions themselves.   In these cases the absorption is probably produced
in the (giant) molecular cloud associated with the HI region.   Deep
features of this type are less common by at least a factor of 10 than
those shown in Figs. 3 and 4.

CO emission line observations around the galactic plane show
a clumpy distribution of these molecules which is mainly confined to an
annulus between galactocentric radii 4 < R < 8 kpc with full thickness
between half density points of ∿100 pc.   There is a considerable discussion
as to whether the molecules are concentrated in "giant" or "large" clouds.
Some observers (for example Solomon & Sanders 1980) have interpreted their
data in terms of giant clouds with a typical density of 300 cm^{-3}, a diameter

Figure 4   The HI, OH and H$_2$CO absorption spectra of the non-
thermal source W49B.   The deepest HI components can be seen in
the molecular spectra although at different relative intensities

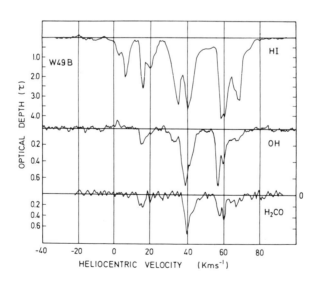

of 40 pc and a mass of 5×10^5 $M_\odot$;   they require 3000-4000 GMCs between
$4 < R < 8$ kpc. On the other hand Burton & Gordon (1978) find that typical
diameters are in the range 5-10 pc and masses are $\geq 2 \times 10^3$ $M_\odot$; 10^6 such
clouds would be required if their diameters were 5 pc. The real situation
is most likely best described by a wider distribution of cloud parameters
than given by either of the models just described. (see for example
Burton & Liszt 1981).

An interesting comparison between HI and CO galactic plane
survey data has been made by Burton, Liszt & Baker (1978) who found that
the molecular emission was strongly correlated with narrow self-absorption
features in the HI profiles. The similarities between the velocities and
the distribution of these two species suggests that residual cold HI
coexists with H_2 in molecular clouds on a galaxy-wide scale similar to
the situation in the dust clouds in the solar neighbourhood.

Concluding comments

A further parameter which is probably important in the
formation of molecular clouds is the magnetic field. There is no doubt
that magnetic fields penetrate both low density HI clouds as well as high
density OH maser clouds which have densities in the range 10^6 to 10^8 cm^{-3}
and magnetic fields with strengths of several milligauss. (See for
example Davies 1981). Fields of this magnitude will have an appreciable
affect on the morphology and collapse timescale of these clouds.

Among the problems relating to interstellar molecules which
requires further investigation is the observation that many molecular
clouds do not appear to have adequate shielding from the interstellar
UV radiation field. For example, the OH, H_2CO and CH molecular clouds
lying in front of Cas A and many other radio sources have a visual extinction at their centres of ≤ 0.5 mag which on most models of molecule
formation is insufficient to protect the molecules against ionization or
dissociation. This result has the following implications. Either the
molecule-formation models are in serious error or the interstellar
clouds are lumpy; the molecules might then be formed preferentially in
the lumps while the HI is mainly concentrated in the space between the
lumps. Most of the correlations described above could probably be
explained on the basis of a lumpy cloud medium. Further high resolution
molecular observations would throw light on this important topic.

References

Baker, P.L. & Burton, W.B., 1975. Astrophys.J. 198, 281.
Batrla, W., Wilson, T.L. & Rahe, J., 1981. Astr.Astrophys. 96, 202.
Bohlin, R.C., Savage, B.D. & Drake, J.F., 1978. Astrophys.J. 224, 132.
Burton, W.B. & Gordon, M.A., 1978. Astr.Astrophys., 63, 7.
Burton, W.B., Liszt, H.S. & Baker, P.L., 1978. Astrophys.J., 219, L67.
Burton, W.B. & Liszt, H.S., 1981. Origin of Cosmic Rays, p227. eds.
        G. Setti, G. Spada & A.W. Wolfendale, Reidel, Dordrecht.
Crovisier, J., 1981. Astr.Astrophys., 94, 162.
Davies, R.D., 1981. Phil.Trans.R.Soc.Lond.A., 303, 581.
Dickman, R.L., 1978. Astrophys.J.Suppl., 37, 407.
Federman, S.R., Glassgold, A.E., Jenkins, E.B. & Shaya, E.J., 1980.
        Astrophys.J., 242, 545.
Federman, S.R., 1981. Astr.Astrophys., 96, 198.
Gordon, M.A. & Burton, W.B., 1976. Astrophys.J., 208, 346.
Heiles, C., 1967. Astrophys.J.Suppl., 15, 97.
Heiles, C. & Gordon, M.A. 1975. Astrophys. J. 199, 361.
Hjalmarson, A., et al. 1977. Astrophys.J.Suppl., 35, 263.
Jenkins, E.B. & Savage, B.D., 1974. Astrophys.J., 187, 243.
Knapp, G.R., 1974. Astronom. J. 79, 527.
Lang, K.R. & Willson, R.F., 1978. Astrophys.J., 224, 125.
Lockhart, I.A. & Goss, W.M., 1978. Astr.Astrophys., 67, 355.
Martin, R.L. & Barrett, A.H., 1978. Astrophys.J.Suppl., 36, 1.
McCutcheon, W.H., Shuter, W.L.H. & Booth, R.S., 1978. Mon.Not.R.astr.
        Soc., 185, 755.
Myers, P.C., 1973. Astrophys.J.Suppl., 26, 83.
Radhakrishnan, V. & Goss, W.M., 1972. Astrophys.J.Suppl., 24, 161.
Radhakrishnan, V., Murray, J.D., Lockhart, P. & Whittle, R.P.J., 1972.
        Astrophys.J.Suppl., 24, 15.
Ryter, C., Cesarsky, C.J. & Audouze, J., 1975. Astrophys.J., 198, 103.
Sandquist, A. & Bernes, C., 1980. Astr.Astrophys., 89, 187.
Savage, B.D., Bohlin, R.C., Drake, J.F. & Budich, W., 1977. Astrophys.J.
        216, 291.
Savage, B.D. & Mathis, J.S., 1979. Ann.Rev.Astr.Astrophys., 17, 73.
Solomon, P.M. & Sanders, D.B., 1980. Molecular Clouds in the Galaxy, p.41.
        eds. P.M. Solomon & M.G. Edmunds, Pergamon, London
Spitzer, L., 1968. Diffuse Matter in Space. Wiley, New York.
van Woerden, H., 1967. Radio Astronomy and the Galactic System, p.3,
        ed. H. van Woerden, Academic Press, London
Wooten, A., Snell, R. & Evans, N.J., 1980. Astrophys.J., 240, 532.

MODELS FOR HOT-CENTRED GALACTIC CLOUDS

M.Rowan-Robinson
Dept. of Applied Maths, Queen Mary College, Mile End Road,
London E1 4NS

Abstract. Theoretical models for the infrared emission
from hot-centred clouds are reviewed. Most work to date
has been on the spherically symmetric case and the
validity of this assumption for the observed clouds
is discussed. Approximations to the equation of radiative
transfer, and their validity, are discussed. A variety
of radiative transfer phenomena have come to light through
accurate numerical solution of the problem of the flow
of radiation through dust and these are reviewed here.
Work on the time-dependent case and on problems with
non-spherical geometry, together with other areas of
development, is also reviewed. Finally some recent
results for individual hot-centred clouds, selected
for their compact size and isolated nature, are presented.
For the BN-type objects (GL490, 989, 4176, 2107, 2591,
and probably S140), a double-shell model, with an inner
shell of hot outflowing dust (perhaps with a bi-polar
structure) surrounded by a second shell of cold dust
(presumably associated with the ambient cloud from which
the young star has formed) is required. For the other
clouds modelled (W3 OH, GL961, GL4176, W75N) a uniform
density cloud illuminated by a hot star gives a good fit
to the observations.

1. THE GEOMETRY OF THE ILLUMINATION OF HOT-CENTRED CLOUDS OF DUST AND MOLECULES

The observational properties of molecular clouds have been
reviewed recently by Evans (1981). The typical giant molecular
cloud is elongated, has a greatest linear extent of 90 pc and has a
mass of 10^5 M$_\odot$. Considerable structure is seen and the complex
line profiles seen in many clouds suggest that this is due to
turbulence. Turbulence can also explain why the lifetime of the
clouds exceeds their free-fall times.

It has been found useful to divide the clouds into
cold clouds (peak gas temperature <20 K) and hot-centred clouds

(T_{peak} > 20 K) (Evans 1978, 1981, Rowan-Robinson 1979). Hot-centred
clouds are usually associated with compact HII regions, infrared
sources and masers, and are currently engaged in the formation of
massive stars. Rowan-Robinson (1979) has defined a relatively
complete sample of hot-centred clouds, based on sources brighter
than -4 magnitudes at 20 µm in the AFGL catalogue.

Before modelling the infrared and submillimetre emission
from these clouds it is essential to consider the geometry of the
cloud and the illuminating star or stars. There are 3 main types
of situation (cf. Rowan-Robinson 1980):

(A) The cloud of dust and molecules is illuminated by
a centrally placed embedded source with a very compact HII region
or no observable HII region, e.g. Mon R2 (Loren 1977), S140 (Blair
et al. 1978) and NGC2244 = GL961 and NGC2264 = GL989 (Blitz 1978).
Even here the geometry may not be spherically symmetric due to
the multiplicity of the illuminating stars (Beichman et al. 1979)
or due to a bi-conical structure of the material near the star
(Harvey et al. 1977a, Rodriguez et al. 1980, Lada & Harvey 1981).

(B) The geometry is highly asymmetric and the cloud
is illuminated by a star in an HII region outside the dense cloud
core, e.g. S255 (Evans et al. 1977), M17 (Gatley et al. 1979), W58
(Israel 1980), G336-0.2 (Hyland et al. 1980), NGC2023 (Harvey et al.
1980).

(C) The cloud is illuminated both by sources embedded within
the dense dust and molecular cloud core and by stars in HII regions
outside the core, e.g. OMC1 (Gehrz et al. 1975, Werner et al. 1976
and refs therein), W75S (Werner et al. 1975, Harvey et al. 1977b
and refs therein), W3 (Hackwell et al. 1978, Tielens & de Jong 1979),
NGC7538 (Werner et al. 1979).

This can be understood if OB stars form in regions of
high density and are at first relatively centrally placed in their
nascent material (class A). However as the HII region evolves and
expands in the direction of least resistance a blister type geometry
develops (Israel 1978, Icke et al. 1980) (class B). Class C
sources occur when a new generation of stars starts to form in the
dense core of a molecular cloud which already has a type B source
to one side of it.

That A-to-B is indeed an evolutionary sequence is strongly suggested by a plot of far infrared surface brightness ($L_{f\ ir}/r_2^2$, where $L_{f\ ir}$ is the total far infrared luminosity and r_2 is the overall radius of the cloud core) against mean molecular hydrogen density n (Fig. 1). Data are taken from Rowan-Robinson (1979): r_2 is defined

Fig 1: The distribution of hot-centred clouds in a plot of far infrared surface brightness versus mean molecular hydrogen density. x's denote clouds of type A, circled points denote clouds of type B, crosses denote clouds of type C. Brackets denote cases where the geometry is probably asymmetric (type B), based on the discrepancy in positions of radio, CO and infrared peaks. Heavy dots are clouds associated with more evolved HII regions of type III or IV (Israel 1976). The long arrow denotes the direction of evolution if the cloud expands keeping $L_{f\ ir}$ and its mass constant. The region marked off to the upper right is where spherically symmetric models may have some validity.

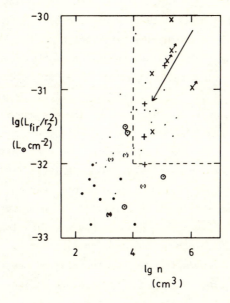

as the radius at which the far infrared intensity drops to 10% of the
peak value. Class A sources are found to the upper right of the
diagram and class B to the lower left. The direction of evolution
in this diagram is from the ultracompact HII regions at the upper
right (type I or II in the classification of Israel 1976) towards the
more extended and evolved HII regions of type III or IV at the lower
left. The sources for which a spherically symmetric geometry may
have some relevance are those with $n > 10^4$ cm^{-3} and $L_{f\ ir}/r_2^2 > 10^{-31}$L cm^{-2}.
The type C sources are also found in this region because the values
of n and r_2 used refer to the molecular cloud core where new
stars are forming.

2. SPHERICALLY SYMMETRIC MODELS FOR HOT-CENTRED CLOUDS

Spherically symmetric models for the infrared emission from
hot-centred clouds have been discussed by a number of authors
(Leung 1976a, Scoville & Kwan 1976, Unno & Kondo 1976, Yorke 1977,
Mitchell & Robinson 1978, Haisch 1979, Rowan-Robinson 1980, 1981,
Yorke & Shustov 1981). Following the ray-tracing method of Hummer
& Rybicki (1971), Leung (1976a,b) has reformulated the equation of
radiative transfer and its moment equations in a form which
renders them susceptible to rapid numerical solution. There are
however three limitations to the accuracy of Leung's work: (i) the
oversimplification of the inner boundary condition, assuming the
radiation field incident on the inner edge of the dust shell to
be isotropic (Mitchell & Robinson 1978, Rowan-Robinson 1980),
(ii) an insufficient number of radial grid points near the inner
edge of the cloud (Rowan-Robinson 1980), (iii) an incorrect boundary
condition for the moment equation, namely that the flux incident
on the inner edge of the cloud is that due to the star
$H_\nu^* = r_s^2 B_\nu(T_s)/4r_1^2$ (where r_s, r_1 are the radii of the star and
the inner edge of the cloud, T_s is the temperature of the star), which
neglects the effect of backwarming on the star (see below, section
4(d)). A generalisation of Leung's approach which corrects these
deficiencies has been developed by Ross Mitchell and some results
will be presented shortly (Mitchell & Rowan-Robinson 1981: see section
4(f) below).

A completely independent exact method of solution of the

equation of radiative transfer has been described by Rowan-Robinson
(1980) and applied to hot-centred dust clouds. Though slower
computationally, this method is mathematically and numerically more
straightforward than the method of Leung and its monitoring of the
accuracy of each step has allowed the problems mentioned above,
together with several of the radiative transfer phenomena discussed
in section 4 below, to come to light. Dirty silicate models have
been discussed for a number of individual sources in Rowan-Robinson
(1981) and some of these results will be summarised in section 6
below.

3.   THE VALIDITY OF VARIOUS APPROXIMATIONS TO RADIATIVE TRANSFER IN DUST

(a) Neglect of scattering

If scattering were entirely in the forward direction
(that is, the phase function for scattering, $\zeta(\theta)$, is a delta-function
and the anisotropy factor, $g = \langle\cos\theta\rangle = 1$), then scattering may be
completely omitted from the problem and we can write the grain
extinction efficiency $Q_{\nu,ext} = Q_{\nu,abs}$. The advantage of this
assumption is that we need calculate only $J_{\nu}(r)$, the zeroth moment of
the intensity $I_{\nu}(r,\theta)$, and can then iterate for the dust temperature
distribution using the assumption of radiative balance. This is the
procedure used by Scoville & Kwan (1976) and it is a useful simplifica-
tion of the problem of radiative transfer in non-spherical geometry
(e.g. axial symmetry, Kandel & Sibille 1978, Icke et al. 1980, Natta
et al. 1981, Walker 1981). Flux constancy may not be very well
satisfied in such an approach.

However reflection nebulae, the zodiacal light, scattered
light from the Milky Way, and scattered light from circumstellar
dust shells, all show that scattering can not be entirely in the
forward direction. The best estimate for the anisotropy factor in
the visual region is $g = 0.7 \pm 0.2$ (Savage & Mathis 1979).

The standard approach in attempts to solve radiative
transfer exactly is to expand $\zeta(\theta)$ in a series of Legendre polynomials
and to retain only the first two terms: $\zeta(\theta) = 1+b_1 \cos\theta$. Since
$g = b_1/3$ and $b_1 < 1$ to ensure $\zeta(\theta) > 0$ for all θ, the maximum value

of g which can be studied by this approach is 1/3 (Rowan-Robinson 1980).
Such models differ little from the case of isotropic scattering and are
poorly represented by the assumption g = 1. By taking further terms
in the series expansion for $\zeta(\theta)$, Mitchell & Rowan-Robinson (1981) have
investigated values of g intermediate between 1/3 and 1.

(b) Include only the first scattering of the starlight

If scattering is fairly isotropic (e.g. g < 1/3) the
most important scattering is the first scattering of the light from
the central star. The scattered light will then have a relatively
isotropic intensity distribution and subsequent scatterings will have
much less effect. The radiated light from grains will be in the
infrared where the scattering efficiency is low.

Rowan-Robinson (1980) has shown that this approximation
gives good results for the dust temperature distribution and for the
emergent infrared flux. Only the first moment of the intensity, $H_\nu(r)$,
need be calculated, and iteration of T(r) proceeds until flux conservation is achieved through the cloud. In this approach, radiative
balance may not be very well satisfied.

(c) Optically thin approximation to get T(r)

This is a good approximation if the ultraviolet optical
depth to the centre of the cloud τ_{uv} < 1, but is a poor one once
τ_{uv} >> 1 because radiation from grains and scattered light then make
an important contribution to the mean intensity, particularly near the
inner boundary of the cloud. The temperature calculated under this
assumption can be off by a factor greater than 2 for τ_{uv} >> 1.

(d) Eddington approximation

The Eddington approximation is to take the variable
Eddington factor $f_\nu = K_\nu/J_\nu$, where K_ν is the second moment of the
intensity, as 1/3 everywhere. One way of doing this is to take the
intensity I_ν to be a linear function of $\mu = \cos\theta$. Alternatively the
radiation field can be considered to be two almost equal and opposite
hemispheres, with isotropic intensities I_ν^+, I_ν^-, and the small
difference between these gives the flux (two-stream approximation).
In either case the Eddington approximation is a very poor one for the

flow of radiation through dust, orders of magnitude off in flux and
not valid at any location or frequency.

Unno & Kondo (1976) have suggested a generalisation in which

$$I_\nu(r,\theta) = I_\nu^+(r) \quad \text{for} \quad 1 > \mu = \cos\theta > \mu_\nu(r),$$

$$= I_\nu^-(r) \quad \text{for} \quad \mu_\nu(r) > \mu > -1,$$

and a version of this has been described by Haisch (1979). The
accuracy of this approximation has yet to be investigated, but
it should be noted that the true intensity distributions calculated
in my exact programme (Rowan-Robinson 1980, Fig. 15) do not all resemble
the above approximation.

(e) Isothermal dust shell

This could only be approximately valid for a geometrically
thin shell and even then there would be temperature structure if
$\tau_{uv} \gg 1$. Far infrared observers have perhaps been attracted to this
assumption by the fact that the observed spectra of hot-centred clouds
can often be fitted by assuming $S_\nu \propto Q_\nu B_\nu(T_d)$ for $\lambda \gtrsim 50\mu m$. The
meaning of T_d is then the mass-weighted average temperature across
the cloud (Rowan-Robinson 1980).

4. RADIATIVE TRANSFER PHENOMENA

(a) Steepening of the temperature gradient at the inner boundary of the cloud

The effect of the back-radiation from other grains and of the
scattered light is to steepen the temperature gradient in the inner part
of the cloud. This is illustrated in Fig. 2. The temperature at
the inner boundary $T(r_1)$ increases by a large factor over the
optically thin value as τ_{uv} becomes $\gg 1$. For fixed $\tau_{uv} > 1$,
increasing the density index β, where $n(r) \propto r^{-\beta}$, has a similar
effect.

(b) The variation of the variable Eddington factor, $f_\nu(r)$

At the inner boundary of the cloud $(r=r_1)$, $f_\nu(r)$ is fixed
by the relative contribution of the star and the radiated and scattered

light from the grains (if the star dominates $f_\nu(r_1) \simeq 1$, corresponding
to a forward-peaked radiation field: if radiation from grains and
scattered light dominate $f_\nu(r_1) \simeq 1/3$, usually slightly less than 1/3

Fig. 2: Temperature distribution for dirty silicate dust cloud
models with $T_s = 2500$ K, $T_1 = 1000$ K, and $\beta = 2$. The right-
hand set of curves are for $r_1/r_2 = 0.05$: the upper solid curve
(at large r) is for $\tau_{uv} < 1$, dotted curve is for $\tau_{uv} = 3$, lower
solid curve is for $\tau_{uv} = 10$, broken curve is for $\tau_{uv} = 30$.
The left-hand set of curves are for $r_1/r_2 = 0.001$ and
$\tau_{uv} < 1$ (upper solid curve), $\tau_{uv} = 10$ (dotted curve), 20
(lower solid curve). Note the steepening of the temperature
gradient at the inner boundary as τ_{uv} increases. Figure
taken from Rowan-Robinson & Harris (1981).

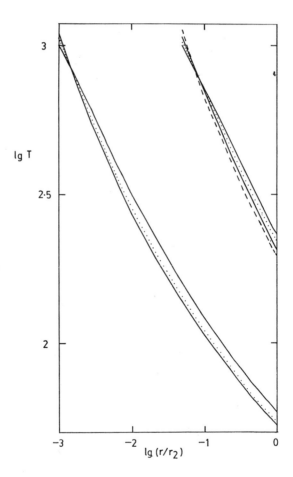

due to sideways beaming - see (c) below). At visible and ultraviolet
wavelengths $f_\nu(r)$ then decreases gradually towards 1/3 as r
increases due to the isotropizing effect of scattering. At near
and middle infrared wavelengths there is a more rapid drop in $f_\nu(r)$
due to the isotropizing effect of radiation from grains: then as r
increases further $f_\nu(r)$ starts to increase because of the anisotropi-
zing effect of the hotter grains being located near the centre of
the cloud. Finally near the outer boundary of the cloud $f_\nu(r)$
increases sharply at all wavelengths (if it is not already $\approx$1)
because the radiation field becomes forward-peaked as it escapes
from the cloud.

(c) Sideways beaming of the radiation field near the inner
     boundary of the cloud

This slightly surprising phenomenon reported by Rowan-
Robinson (1980) is a geometrical effect due to the fact that the

Fig. 3: Comparison of a portion of the (blackbody) spectrum
radiated by the star (dotted curve) with that entering the
dust cloud (solid curve). The latter is slightly reduced,
especially at wavelengths near 10 and 20 μm, due to the
backwarming of the star by the cloud. The model shown
here has T_S = 2500 K: the effect is even smaller if
T_S = 40,000 K.

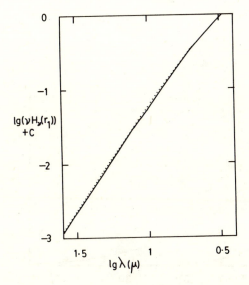

hottest grains have an almost planar distribution locally at $r = r_1$.
It occurs whether radiation from grains or scattered light is the
dominant contribution to the non-stellar light. A consequence is
that, if the stellar contribution to the mean intensity is negligible,
the variable Eddington factor $f_\nu(r)$ takes values $< 1/3$.

(d) Backwarming of the star

If nuclear energy sources provide a luminosity $L = 4\pi r_s^2 \sigma T_s^4$,
then inside a circumstellar dust cloud, the cloud will radiate an
additional amount ΔL back at the star. The stellar temperature will
adjust to $T_s' = T_s + \Delta T_s$, where $L + \Delta L = 4\pi r_s^2 \sigma T_s'^4$ (actually the
whole photospheric structure will be altered). The net flux entering
the dust cloud will still be L, but it does not have a blackbody
spectrum. In fact the spectrum will have the form: $H_\nu^* = r_s^2 B_\nu(T_s')/4r_1^2$
- backwarming flux. This is illustrated in Fig. 3. Another way of
seeing this is to consider equal and opposite cones centred on a point

Fig. 4: Emergent intensity profiles at various wavelengths,
relative to the intensity at the rim of the cavity, for
a model for W3 OH with parameters: T_s = 40,000 K,
$T_{1'}$ = 300 K, r_1/r_2 = 0.01, β = 0, τ_{uv} = 200.

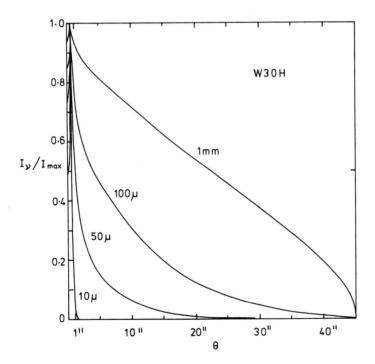

at $r = r_1$. If the cones do not include the star then they will both
see the same flux and there is no net contribution to H *. However,
the cone subtended by the star does not receive the radiation from
the opposite side of the cloud (shadowed by the star), which
would have cancelled the contribution from the opposite cone
pointing away from the star. The larger r_s/r_1, and hence the angle
of this latter cone, the more significant will be the effect of
backwarming.

    In my radiative transfer programme (Rowan-Robinson 1980)
I keep T_s' fixed and allow the model to iterate towards a lower value
of L. Codes like that of Leung (1976a,b), in which both T_s' and H *
are held fixed, are incorrect, though this is not very serious
for hot stars.

(e) Emergent intensity profiles

    Fig. 4 illustrates the predicted emergent intensity profiles
for a dirty silicate grain model for W3 OH (Rowan-Robinson 1981 and
section 6 below). The characteristic feature of these profiles is
the bright annulus of radiation predicted at the rim of the cavity.

   Fig. 5: Comparison of emergent spectra for isotropic (solid
   curve) and completely forward scattering (dotted curve) for
   model with T_s = 40,000 K, T_1 = 1000 K, τ_{UV} = 10, β = 0,
   r_s/r_2 = 5 × 10^{-7}, and grains for which $Q_{\nu,abs}$ = $4\pi a/\lambda$,
   a = 0.1 μm (Rowan-Robinson 1980).

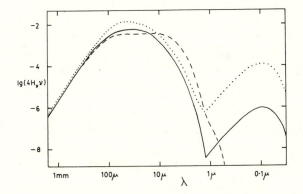

Other examples of predicted intensity profiles for hot-centred clouds
are given in Leung (1976a), Yorke (1977), Mitchell & Robinson (1978),
Rowan-Robinson (1980, 1981).

(f) Effects of scattering

As the anisotropy factor g is reduced from 1 (completely
forward scattering) to 0 (isotropic scattering) there is a drop of
$\exp(-a_\nu \tau_\nu)$ in the transmitted light from the star, where a_ν is the
albedo. This is partially compensated for by the contribution of
scattered light to the emergent flux. Fig. 5, taken from Rowan-
Robinson (1980), illustrates this effect. There is also a change in
the temperature distribution of the grains, it becoming hotter
near the centre and colder near the edge of the cloud. The detailed
dependence of the cloud structure on the anisotropy factor g will
be discussed by Mitchell & Rowan-Robinson (1981). For g < 1/3 there
is little difference from the isotropic case (Rowan-Robinson 1980).

There is also a dramatic increase predicted in the apparent
size of the source at visible and ultraviolet wavelengths due to
scattering as g changes from 1 to 0. For high optical depth ($\tau_{uv} \gg 1$)
scattering also begins to increase the size of the source at near
infrared wavelengths.

(g) Line profiles

A question which has been much discussed is: under what
conditions do the 10 and 20 µm silicate features appear in absorption or
emission? Clearly at very low optical depths the features must appear
in emission while at very high optical depths the features must appear in
absorption. Where the change-over occurs depends on the relative optical
depths in hot and cold grains in the cloud and for $0 < \beta < 2$, the more
extended the cloud the more likely a line is to be in absorption (Rowan-
Robinson 1980, Mitchell & Robinson 1981, Yorke & Shuster 1981). Mitchell
& Robinson (1981) have given a very detailed discussion of the conditions
for line formation in absorption or emission, and have pointed out that
at intermediate optical depths, self-absorption would occur (i.e., wings
in emission, core in absorption).

(h) Dynamical effects of a radiation pressure gradient

Each photon carries a momentum $h\nu/c$, so the absorption and scattering (provided $g \neq 1$) of photons transfers momentum to the cloud. The total radial momentum absorbed per sec is $\tau_H L/c$, where τ_H is the effective (i.e., using the effective extinction efficiency $Q_{\nu,abs} + (1-g) Q_{\nu,sc}$) optical depth at the peak of the flux $H_\nu(r)$ averaged over the different zones of the cloud (Rowan-Robinson 1981, Appendix). τ_H is a number of order $\tau_{\bar{\nu}}$ where $\bar{\nu}$ is the peak frequency of the input spectrum from the star. Note that τ_H may be much larger than unity, even if scattering is not very significant.

5. PROBLEMS AND AREAS OF DEVELOPMENT

(a) Specification of the grain temperature at the inner boundary

In reality the temperature at the inner boundary of the cloud, $T(r_1)$, is presumably determined by the condensation temperature of the grains in most cases. In practice we have to specify r_1/r_s before we start a radiative transfer calculation: we can define this through T_1, the temperature the grains would have if illuminated only by the star. We then have to iterate to find the true value of $T(r_1)$ (Rowan-Robinson 1980). To try to specify $T(r_1)$ and then iterate to find r_1 would be computationally impractical at present.

(b) Specification of the stellar temperature

The parameter that is determined by stellar evolution theory is T_s, the surface temperature the star would have in the absence of the dust cloud. The presence of the dust cloud raises this to T_s' due to the effect of backwarming (section 4(d) above). In practice it is inconvenient to specify the flux entering the cloud and iterate allowing the stellar temperature to change. It is more convenient to specify T_s' and allow the flux entering the cloud to iterate to a lower value. The effect is small for the hot stars of interest in this review.

(c) The time-dependent case

Time-dependent hydrodynamic calculations of the formation of massive stars from molecular clouds have been carried out by Yorke &

Krügel (1975, 1977) and the resulting infrared spectra have been calculated for several phases by Yorke (1977). Fig. 6 shows the
density distribution calculated immediately preceding the formation of a
compact HII region and the resulting infrared spectrum for silicate-ice
grains. Although an n(r) ∝ r^{-2} density distribution of hot dust
surrounded by a second shell of colder dust does indeed give a good
fit to the observed spectra of BN type objects (see section 6 below),
the evidence from molecular line observations favours outflow rather
than infall (e.g. Solomon et al. 1981, Beckwith 1981) and the shell of
cold dust has a more direct explanation as the ambient cloud from
which the star has formed. Nevertheless hydrodynamic calculations,
particularly in which the radiative transfer is treated accurately,
are of the greatest importance.

Fig. 6: The density profile calculated for a protostellar
cloud just before the formation of a compact HII region by Yorke
& Krügel (1977), and the corresponding emergent spectrum for
silicate-ice grains calculated by Yorke (1977).

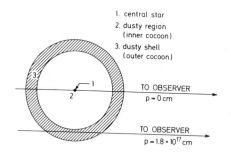

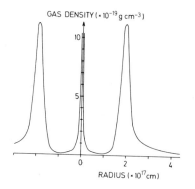

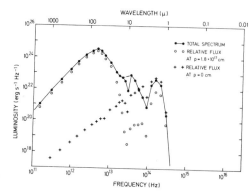

(d) Non-spherical geometry

As discussed in section 1 there is ample observational
evidence for departures from spherical symmetry in hot-centred clouds.
Unfortunately the mathematical and computational difficulties in the way
of developing a code which can solve the equation of radiative transfer
in dust with an arbitrary spatial distribution are formidable. Even
where the distribution of dust is taken to be axially symmetric (i.e.,
2-dimensional), the ray geometry still requires a full 3-dimensional
treatment. Icke et al. (1980) have simplified the problem still further
in their treatment of a hot star in a density gradient. As well as
neglecting scattering, they assume that photons remain in the cone in
which they are emitted and that the radiation field can be treated as
the sum of 3 delta-functions, corresponding to Lyman continuum,
Lyman α and softer radiation. Moreover the contribution of radiation
from dust to the mean intensity is neglected in calculating the dust
temperature and the grains are assumed to have a power-law absorption
efficiency at all wavelengths. With such drastic assumptions their
results can be of qualitative interest only. Earlier Kandel &
Sibille (1978) treated flattened density distributions in a similar way.

Natta et al. (1981) have discussed the problem of a hot
star located near the edge of a homogeneous cloud. While they too
neglect scattering, they do calculate the mean intensity directly using
a 13 x 13 x 10 grid in (r, θ, ϕ). An interesting result they find is
that the dust temperature is fairly symmetric about the source. Walker
(1981) has studied a similar problem with a more general density
distribution for the cloud, though to date he has included only the
radiation from the star in calculating the dust temperature.

Although considerable progress may be expected for the case
where scattering is neglected, the role of scattering is of the
greatest interest in reflection nebulae and related phenomena of
non-spherical geometries.

(e) Several grain species

The importance of multiple grain species has been emphasized
by Sarazin (1978), Mitchell & Robinson (1978) and Tielens & de
Jong (1979). Leung (1976b) has described how multiple grain species
can be incorporated into his approach. Haisch (1979) has presented
some results for multiple grain species, based on the modified

Eddington assumption approach of Unno & Kondo (1976) - but see section 3(d) above.

(f) The high albedo case

Rowan-Robinson (1980) has pointed out the difficulty of solving the high albedo case ($Q_{\nu,sc} \gg Q_{\nu,abs}$) with methods which iterate on the dust temperature distribution, since the latter has little effect on the radiation field in this case. If this situation proves to be of practical importance (as it would be for terrestrial or lunar silicates, for example), a Monte Carlo type of approach may be necessary.

(g) The stability of convergence of different methods

Results from Leung-type programmes developed independently by Ian Butchart, of Preston Polytechnic, and Ross Mitchell, of Queen Mary College, have now been compared in detail with my programme (Rowan-Robinson 1980). After correction of the problems mentioned in section 2 above, the agreement is excellent, to better than 1% in the temperatures and variable Eddington factors. However, there are regions of parameter space where each programme fails to converge, for values of $\tau_{uv} \gg 100$ for example. The reasons for these failures and the extent to which they can be overcome has yet to be investigated.

6.   RESULTS FOR INDIVIDUAL HOT-CENTRED CLOUDS

Figs 7a, b show dirty silicate model fits to the infrared spectra of a number of individual sources, selected for their compact size and isolated nature, so that they are the most likely to satisfy the assumption of spherically symmetric illumination by s single star (taken from Rowan-Robinson 1981). In each case the star is assumed to have a temperature of T_s = 40,000 K and the temperature of the hottest grains is taken to be $T_1 \doteq$ 1000 K, unless otherwise specified.

For W3 OH, GL961, GL4182, W75N, a uniform density cloud illuminated by an early-type star gives a reasonable fit to the observed spectra. This suggests that the molecular clouds

Fig. 7a: Observed spectra for selected clouds, $\log_{10} (\nu S_\nu) + C$, in ergs s^{-1} cm^{-2}, compared with theoretical models (Rowan-Robinson 1981). In each case T_s = 40,000 K, T_1 = 1000 K, r_1/r_2 = 0.001, unless otherwise specified.
GL331 = W3OH: C = 7, β = 0, τ_{uv} = 200 (dotted curve); broken
  curve: same but T_1 = 300 K, r_1/r_2 = 0.01
GL961 = N2244: C = 4, β = 0, τ_{uv} = 40.
GL2621 = W75N: C = 1.5, β = 0, τ_{uv} = 200, r_1/r_2 = 0.00316.
GL2884 = S140: C = -2, broken curve: β = 0, τ_{uv} = 100;
  dotted curve:β= 1, τ_{uv} = 150; solid curves: double-
  shell model.

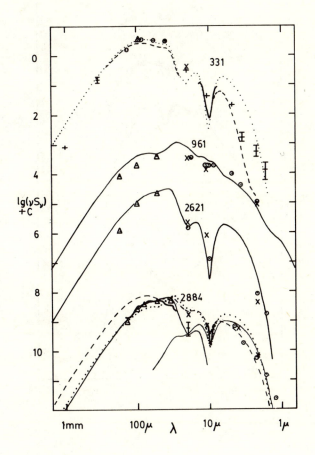

Fig. 7b: GL490 double-shell model, dotted curve: $\beta = 2$,
$\tau_{UV} = 40$, $\tau^C_{10\mu} = 1$; solid curve, $\beta = 0$.
GL989 = N2264: broken curve: $\beta = 1$, $\tau_{UV} = 100$; dotted curve:
$\beta = 2$ (double-shell), $\tau^C_{10\mu} = 0.5$; solid curve, $\beta = 0$.
GL2591: solid curve: $\beta = 2$(double-shell), $\tau_{UV} = 100$, $\tau^C_{10\mu} = 2$;
dotted curve: same but $T_s = 2500$ K.
GL4176: $\beta = 2$(double-shell), $\tau_{UV} = 50$, $\tau^C_{10\mu} = 3$.
GL4182: $\beta = 0$, $\tau_{UV} = 100$.

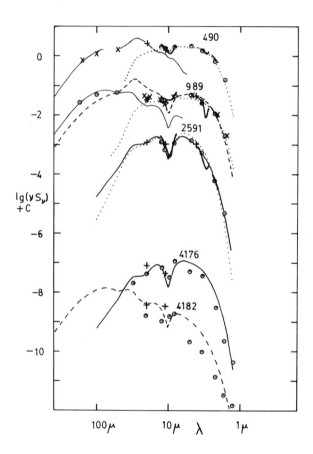

from which the stars have formed are not in a state of collapse. For W3 OH, a model with T_1 = 300 K is found to give a good fit to the observations at λ > 3 μm and the angular radius of the inner edge of the dust shell then agrees well with that of the shell of ionised gas seen by Dreher & Welch (1981) and Scott (1981) at θ = 0".5 - 0".8. The model does not predict enough radiation at 1.6 and 2.2 μm and this could be due either to an overestimate of the grain absorption efficiency by a factor of ∿2 at these wavelengths or to emission from some hot dust within the HII region. This clearing of (most of) the dust from a zone ten times larger than the grain melting radius suggests that W3 OH has evolved slightly further than the other sources and this is consistent with its higher position in the log L - log($N_e^2 V$) diagram of Thompson (1981), between the zero-age main-sequence and supergiant loci.

For the BN-type objects (GL490, 989, 4176, 2107, 2591) the spectrum does not seem to be consistent with a single power-law density distribution. For these a double-shell model, with an inner shell of hot dust in which the density $n(r) \propto r^{-2}$ and a second shell of cold absorbing dust (optical depth at 10 μm, $\tau_{10\mu}^c$), presumably associated with the ambient cloud, is fitted. This model gives a good fit to the $\lambda \lesssim$ 10-20 μm observations of these sources, in several of which a high-velocity outflow has been observed (Rodriguez et al. 1980, Snell et al. 1980, Solomon et al. 1981, Lada & Harvey 1981, Beckwith 1981).

To explain the excess radiation observed at λ > 20 μm in GL490 and 989, there are two simple possibilities. If the assumption of spherical symmetry is broadly correct then the far infrared radiation could simply be re-emission of the radiation absorbed by the cold ambient dust cloud from the hot outflowing dust. Alternatively, since CO observations suggest a bi-polar structure (Lada & Harvey 1981), a second possibility is that the hot outflowing $n(r) \propto r^{-2}$ dust is confined to a bi-conical region of semi-angle α, absorbing a fraction 1 - cosα of the star's output. The remainder would be absorbed directly by the ambient cloud and be re-emitted in the far infrared. Both these models give a good fit to the observed far infrared spectra. A sketch of the second model is given in Fig. 8.

The overall spectrum of S140 can be fitted by an $n(r) \propto r^{-1}$ density distribution, but a 2-component model similar to the above may be more plausible.

The masses of hot outflowing dust and gas inferred from the above models for BN-type objects lie far below those estimated from CO, e.g. for GL490 I estimate M_{gas} = 0.034 $M_\odot$ in hot outflowing gas (assuming M_{dust} = 0.02 M_{gas}, V_{exp} = 25 km s^{-1}, and r_2 = 0.29 pc), compared with 16-45 $M_\odot$ estimated from the CO outflow (Lada & Harvey 1981). Although the masses of gas inferred from the above models for the infrared emission could be accelerated by radiation pressure acting on grains, the CO observations, if correctly interpreted, imply that a substantial fraction of the ambient cloud has been accelerated to high velocity and that the acceleration mechanism can not be radiation pressure acting on dust. These sources clearly deserve further theoretical and observational study.

Fig. 8: Sketch of possible configuration for a bi-polar outflow from a star embedded in an ambient molecular cloud.

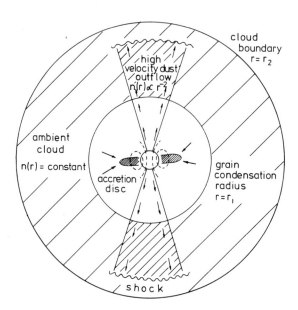

References

Beckwith, S. (1981). IAU Symposium No. 96, Infrared Astronomy, ed. C. G.
        Wynn-Williams & D. P. Cruikshank, P.167. Dordrecht: Reidel.
Beichman, C. A., Becklin, E. E., & Wynn-Williams, C. G. (1979).
        Astrophys. J. 232, L47.
Blair, G. N., Evans, N. J. II, Becklin, E. E., & Neugebauer, G. (1978).
        Astrophys. J. 219, 896.
Blitz, L. (1978). Ph.D. dissertation, Columbia University.
Dreher, J. W. & Welch, W. J. (1981).   Astrophys. J. 245, 857.
Evans, N. J. II (1978). In Protostars and Planets, ed. T. Gehrels, p.153.
        Tucson: Univ. of Arizona Press.
Evans, N. J. II (1981). IAU Symposium No. 96, Infrared Astronomy,
        ed. C. G. Wynn-Williams & D. P. Cruikshank, p. 107.
        Dordrecht: Reidel.
Evans, NJ. II, Blair, G. N., & Beckwith, S. (1977), Astrophys. J.
        217, 729.
Gatley, I., Becklin, E. E., Sellgren, K., & Werner, M. W. (1979).
        Astrophys. J. 233, 575.
Gehrz, R. D., Hackwell, J. A., & Smith, J. R. (1975). Astrophys. J.
        202, L33.
Hackwell, J. A., Gehrz, R. D., Smith, J. R., & Briotta, D. A. (1978).
        Astrophys. J. 221, 797.
Haisch, B. M. (1979). Astron. Astrophys. 72, 161.
Harvey, P. M., Campbell, M. F., & Hoffmann, W. F. (1977a).
        Astrophys. J. 215, 151.
Harvey, P. M., Campbell, M. F., & Hoffmann, W. F. (1977b).
        Astrophys. J. 211, 786.
Harvey, P. M., Campbell, M. F., & Hoffmann, W. F. (1980). Astrophys. J.
        235, 894.
Hummer, D. G., & Rybicki, G. B. (1971). Mon. Not. R. astr. Soc. 152, 1.
Hyland, A. R., McGregor, P. J., Robinson, J., Thomas, J. A., Becklin,
        E. E., Gatley, I., & Werner, M. W. (1980). Astrophys. J.
        241, 709.
Icke, V., Gatley, I., & Israel, F. (1980). Astrophys. J. 236, 808.
Israel, F. (1976). Ph.D. dissertation, Leiden University.
Israel, F. (1978). Astron. Astrophys. 70, 769.
Israel, F. (1980). Astrophys. J. 236, 465.
Kandel, R. S., Sibille, F. (1978). Astron. Astrophys. 68, 217.
Lada, C. J., Harvey, P. M. (1981). Astrophys. J. 245, 58.
Leung, C. M., (1976a). Astrophys. J. 209, 75.
Leung, C. M. (1976b). J. Quant. Spectrosc. and Rad. Transfer 16, 559.
Loren, R. B. (1977). Astrophys. J. 215, 129.
Mitchell, R. M., & Robinson, G. (1978). Astrophys. J. 220, 841.
Mitchell, R. M., & Robinson, G. (1981). Mon. Not. R. astr. Soc. 196,801.
Mitchell, R. M., & Rowan-Robinson, M. (1981) in preparation.
Natta, A., Palla. F., Panagia, N., & Preile-Martinez, A. (1981).
        Astron. Astrophys. 99, 289.
Rodriguez, L. F., Ho, P. T. P., & Moran, J. M. (1980). Astrophys.
        J. 240, L149.
Rowan-Robinson, M. (1979). Astrophys. J. 234, 111.
Rowan-Robinson, M. (1980). Astrophys. J. Supp. 44, 403.
Rowan-Robinson, M. (1981). Mon. Not. R. astr. Soc. (in press).
Rowan-Robinson, M., & Harris, S. (1981). Mon. Not. R. astr. Soc.
        (in press).
Sarazin, C. L. (1978). Astrophys. J. 220, 165.

Savage, B. D., & Mathis, J. S. (1979). Ann. Rev. Astron. Astrophys.17,73.
Scott, P. F. (1981). Mon. Not. R. astr. Soc. 194, 25P.
Scoville, N. Z., & Kwan, J. (1976). Astrophys. J. 206, 718.
Snell, R. L., Loren, R. B., & Plambeck, R. L. (1980). Astrophys. J.
         239, L17.
Solomon, P. M., Huguenin, G. R., & Scoville, N. Z. (1981). Astrophys.
         J. 245, L19.
Thompson, R. I. (1981). IAU Symposium No. 96, Infrared Astronomy,
         ed. C. G. Wynn-Williams & D. P. Cruikshank, p.153.
         Dordrecht: Reidel.
Tielens, A. G. G. M., & de Jong, T. (1979). Astron. Astrophys. 75, 326.
Unno, W., & Kondo, M. (1976). Publ. Astron. Soc. Japan 28, 374.
Walker, D. (1981). This volume.
Werner, M. W., Elias, J. H., Gezari, D. Y., Hauser, M. G., &
         Westbrook, W. E. (1975). Astrophys. J. 199, L185.
Werner, M. W., Gatley, I., Harper, D. A., Becklin, E. E., Loewenstein,
         R. F., Telesco, C. M., & Thronson, H. A. (1976).
         Astrophys. J. 204, 420.
Werner, M. W., Becklin, E. E., Gatley, I., Matthews, K., Neugebauer, G.,
         & Wynn-Williams, C. G. (1979), Mon. Not. R. astr. Soc.
         188, 463.
Yorke, H. W. (1977). Astron. Astrophys. 58, 423.
Yorke, H. W., & Krügel, E. (1975). Mitt. Astron. Ges. 38, 222.
Yorke, H. W., & Krügel, E. (1977). Astron. Astrophys. 54, 183.
Yorke, H. W., & Shustov, B. M. (1981). Astron. Astrophys. **98, 125.**

MODELS OF EXTERNALLY HEATED CLOUDS: AN INTERIM REPORT.

David W. Walker
Physics Department, Queen Mary College, Mile End
Road, London E1 4NS.

Abstract. Models of externally heated clouds are presented. The dust temperature distribution is evaluated, and used to find the model surface brightness distribution, energy spectrum and colour temperature. A qualitative comparison is made between one of the models and infrared observations of M17.

Introduction

The main impetus behind this work is to determine whether infrared and molecular line observations of molecular clouds necessarily require a heat source within the cloud, or whether, in some cases, the observations may be accounted for by models in which the cloud is heated externally.

In the models presented here the dust temperature distribution and the infrared emission of the cloud are evaluated; however, I intend to develop the model further to determine the kinetic temperature of the gas, on the assumption that the gas is heated primarily by collisions with the dust, and hence to determine the expected molecular line emission. This more developed model will then be used to try to account for the recent observations of M17SW by Beckman and co-workers (Chown et al. and Beckman et al. (1981)), who found that the positions of the peak emission in the $C^{12}O(J = 1{\rightarrow}0)$ and $HNC(J = 3{\rightarrow}2)$ lines at 115 GHz and 272 GHz, respectively, are not coincident. This strongly suggests that spherically symmetric models are not applicable in this case.

Since this is essentially a work-in-progress report rather than the presentation of finalised models, no attempt

will be made to quantitatively fit models to observations.
However, a qualitative comparison will be made with
observations of M17 by Gatley et al. (1979).

The Model

We consider a spherical cloud of dust and gas of
radius R, heated by a star at distance $\delta(>R)$ from the cloud
centre. The density distribution of the cloud is assumed
to be spherically symmetric about the cloud centre, and
thus the temperature distribution of the dust and gas will
be symmetric with respect to rotation about a line passing
through the cloud centre and the star. This line will be
referred to as the axis of symmetry.

The position of some general point, P, within the
cloud will be specified by polar co-ordinates (r,θ,ϕ), and
the distance between P and the star will be denoted by s,
as shown in figure 1.

In the evaluation of the dust temperature distribution, $T(r,\theta)$, the presence of the gas may be ignored
in the thermal balance of the dust. In addition, the

Figure 1. The geometry of the model.

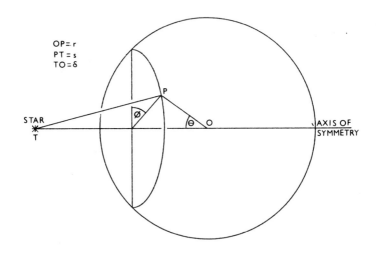

OP = r
PT = s
TO = δ

STAR

T

P

ϕ

θ O

AXIS OF
SYMMETRY

following assumptions are made:

     (i) Scattering by dust grains is ignored.

   (ii) The possible formation of an HII region on
         the side of the cloud nearest the star is ignored.

 (iii) the absorption efficiency of the grains is unity
         for photons emitted by the star, and is given by:

$$Q(\lambda) = 1/\lambda(\mu m) \tag{1}$$

      for photons emitted by the dust.

      The first assumption is justified since the star
emits most of its energy at short wavelengths, for which
the grains are strongly forward-scattering. The grains,
however, emit predominantly at infrared wavelengths, where
the scattering efficiency is low.

      If the star emits sufficiently strongly in the
Lyman continuum, and is close enough to the cloud, an
HII region will form near the edge of the cloud. If this
HII region is small compared with the radius of the cloud,
it will have only a small effect upon the far infrared
emission of the cloud. Since any HII region will absorb all
the Lyman continuum photons emitted by the star, it will be
assumed that the dust is not heated by stellar photons above
the Lyman limit.

      The mathematical formulation of the problem of
determining the dust temperature distribution has been
described by Natta et al. (1981), who considered a cloud of
constant density heated by a star within the cloud. Only
a brief description of the mathematics will be given here.

      The dust temperature distribution is determined
by the equation of radiative equilibrium:

$$4\pi a^2 \int_0^\infty \pi Q(\lambda) B_\lambda \left[T(r,\theta)\right] d\lambda = \pi a^2 \int_0^\infty Q(\lambda) F_\lambda(r,\theta) d\lambda \tag{2}$$

where a is the grain radius, and $F_\lambda(r,\theta)$ is the flux of
radiation of wavelength λ at (r,θ).

      Performing the integration in equation 2, we have:

$$CT(r,\theta)^5 = F_*(r,\theta) + F_{IR}(r,\theta) \qquad (3)$$

where $C = 6 \times 10^{-8} \text{erg cm}^{-2}\text{deg}^{-5}\text{s}^{-1}$. $F_*(r,\theta)$ and $F_{IR}(r,\theta)$ are fluxes due to stellar and infrared emission, respectively, and are given by:

$$F_*(r,\theta) = B_* \frac{\exp(-\tau)}{s^2} \qquad (4)$$

$$F_{IR}(r,\theta) = 3.2 \times 10^{-12}\tau_{uv} \int_0^1 \int_0^\pi \int_0^\pi n(r')(r')^2 \text{Sin}\theta' \; T(r',\theta')^6 .$$

$$\frac{\zeta(6,q)}{d^2} \cdot d\phi' d\theta' dr' \qquad (5)$$

where

$$B_* = \frac{L_*}{4\pi R^2}$$

$$\zeta(m,q) = \sum_{n=1}^{\infty} (n+q)^{-m}$$

$$q = 6.95 \times 10^{-5} T(r',\theta') \cdot t(r,\theta;r',\theta',\phi') \qquad (6)$$

$$\tau_{uv} = \pi a^2 n_o R .$$

In equation 4, τ is the dust optical distance between the star and (r,θ). In equation 5, d is the geometrical distance between the points (r,θ) and (r',θ',ϕ') within the cloud, whilst t in equation 6 is the optical distance between the same points if the absorption efficiency is set equal to unity.

All geometrical distances in equations 4 - 6 are expressed as fractions of the cloud radius, R, and n(r) is the dust density at radial distance r, expressed as a fraction of the central density, n_o.

Since the dust temperature appears in the expression for $F_{IR}(r,\theta)$, the dust temperature distribution can only be found by an iterative process. In all models presented in this work a first approximation to $T(r,\theta)$ will be found by setting $F_{IR}(r,\theta)$ in equation 3 equal to zero.

Thus the heating due to the infrared emission of the dust
will be ignored. This considerably simplifies the problem
although it is intended to include the heating due to the
dust in later models.

In all the models considered here the density
distribution is assumed to be Gaussian, with a half-width
at half maximum chosen so that the density at the surface
of the cloud is 0.001 n_o.

The dust temperature distribution is then determined
by just three parameters:

   (i) $B_* = L_*/4\pi R^2$, where L_* is the luminosity of
       the star below the Lyman limit,

  (ii) δ, the distance of the star from the cloud centre,

 (iii) τ_* the optical thickness of the cloud for stellar
       photons.

To determine the infrared surface brightness
distribution of the model source we must also specify:

  (iv) ω, the orientation angle. This is the angle
       between the axis of symmetry and a line in the
       direction of the observer passing through the
       cloud centre.

Results of the Models.

We first illustrate the effect of varying the
optical depth parameter, τ_*, on the dust temperature distribution. Figure 2 shows the dust temperature along the
axis of symmetry for a number of values of τ_*. In all cases
$B_* = 10$ erg cm^{-2}s^{-1} and $\delta = 1.5$.

For small values of τ_* the attenuation factor in
the expression for $F_*(r,\theta)$ is unimportant so that $T \propto s^{-0.4}$;
however, as τ_* increases the attenuation of stellar photons
becomes more important and the dust temperature at a given
point decreases. It should be remembered that, as τ_*
increases, the heating of the dust by infrared photons from
other dust grains becomes important at large optical distances
from the star. To illustrate this, and to estimate the
minimum value of τ_* for which heating by infrared photons
becomes important, an estimate of the upper limit to $F_{IR}(r,\theta)$

may be obtained by setting t in equation 6 equal to zero.
Figure 3 shows the relative heating, $G(r,\theta) = F_*(r,\theta)/F_{IR}(r,\theta)$,
along the axis of symmetry for a number of values of τ_*.
Again $B_* = 10$ erg cm^{-2}s^{-1} and $\delta = 1.5$ for all models.

As expected, the heating by dust grains becomes
more important as τ_* increases; however, figure 3 shows
that it may be ignored for $\tau_* \lesssim 5$.

We next consider in more detail two models, one
with $\tau_* = 0.1$ (model 1), and the other with $\tau_* = 250$ (model 2).
For both models the following parameters were adopted:

$$B_* = 10 \text{ erg cm}^{-2}\text{s}^{-1}$$
$$\delta = 1.5$$
$$\omega = 90^{\circ}.$$

In model 1 it has already been shown that it is
valid to ignore the heating by infrared photons, although in
model 2 these photons dominate the heating over part of the
cloud. Even so, the heating in most of this region is

Figure 2. The variation of dust temperature along
the axis of symmetry for different values of τ_*.

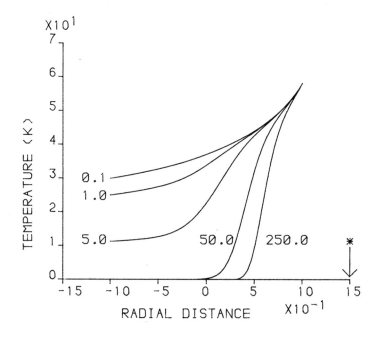

Figure 3. Relative heating, G, along the axis of symmetry for different values of τ_*.

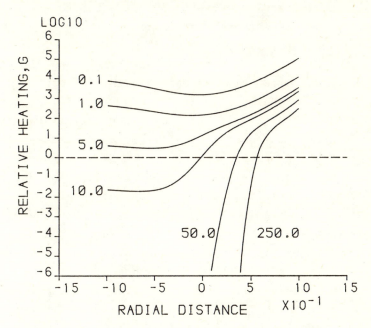

Figure 4. Surface brightness distribution at 30 μm for (a) model 1, and (b) model 2. For model (a) maximum and minimum contour levels are 1.75 and 0.25 x 10^{-16}erg $cm^{-2}s^{-1}$ sterad^{-1}Hz^{-1}, respectively, whilst for (b) they are 500 and 100 x 10^{-16}erg $cm^{-2}s^{-1}$ sterad^{-1} Hz^{-1}. Dashed line indicates the boundary of the cloud.

(a)   (b)

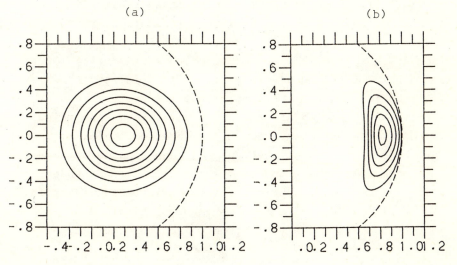

sufficiently small to ensure that the dust temperature does not rise above a few degrees K, so that the omission of the heating by infrared photons is not expected to affect the model surface brightness distribution, or the energy spectrum for wavelengths $\lesssim$ 1 mm.

Figure 4 shows the surface brightness distribution of each model at 30 μm. As can be seen, the contours for model 1 are generally circular, whilst in model 2 they are considerably elongated. The reason for this is that the temperature of the dust in model 1 is approximately constant over much of the cloud so that the surface brightness distribution, particularly at far infrared wavelengths, tends to follow the density distribution. For model 2, the dust temperature is significantly large only within the rim of the cloud nearest the star, and it is the dust temperature distribution within this rim which largely determines the infrared surface brightness distribution.

The variation of the position of the peak of the surface brightness distribution as a function of wavelength has also been investigated. It was found that in both models the position of the peak was nearer to the cloud centre at longer wavelengths, as shown by figure 5.

This is to be expected since the cooler grains are responsible for most of the emission at longer wavelengths. For model 1, in which the surface brightness distribution is largely determined by the density distribution, the distance, r_p, of the peak from the cloud centre falls rapidly with increasing wavelength, and is close to zero for $\lambda \gtrsim$ 100 μm. For model 2 the decrease in r_p is less rapid, and r_p never falls below $\sim$0.6 for λ < 1 mm, since the grains are not significantly heated at radial distances less than this.

The effect of varying orientation angle, ω, upon the shape of the model spectrum has been considered. In the case of model 1 the shape of the spectrum is independent of ω since the cloud is optically thin to infrared radiation, and we observe emission from throughout the cloud for all values of ω. As ω increases from $0°$ to $180°$ the flux at

shorter wavelengths decreases, and the spectrum becomes
narrower. This is because the emission at short infrared
wavelengths is attenuated by the cloud, whereas at longer
wavelengths it is not. Hence, as ω increases we see less of
the short wavelength radiation emitted by the high temperature
rim of the cloud.

The 50-100 μm colour temperature was found to be
constant over most of the cloud in model 1, but showed a
strong decrease towards the centre of the cloud in model 2.
Again, this is because the surface brightness distribution
in model 1 tends to follow the density distribution whereas
in model 2 the rapid decrease in the dust temperature as
we move from the edge of the cloud near the star towards the
cloud centre, gives rise to the strong gradient in the colour
temperature.

Figure 5. Variation of distance of position of
peak intensity from cloud centre with wavelength.

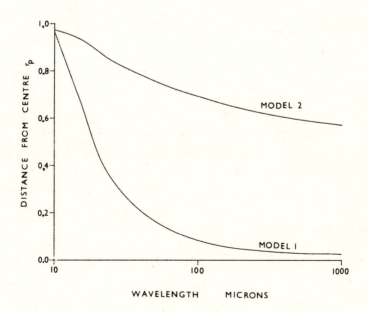

Conclusions

The results of the preceding section may be summarised as follows. For model 1:

  (i) $T(r,\theta) \propto L_*^{0.2} s^{-0.4}$

 (ii) the infrared surface brightness distribution shows little deviation from circular symmetry and tends to follow the density distribution, particularly for wavelengths $\gtrsim 100$ μm

(iii) the shape of the spectrum is independent of the orientation angle, ω.

In the case of model 2 we may draw the following conclusions:

1. high temperature rim on the side of the cloud nearest to the star,

2. the infrared surface brightness distribution is elongated,

3. the shape of the spectrum depends on the orientation angle, ω,

4. there is a strong gradient in the 50 - 100 μm colour temperature.

In both models the distance between the position of the peak surface brightness and the cloud centre decreases with increasing wavelength.

It is of interest to compare the results for model 2 with the infrared observations of M17 by Gatley et al. (1979). Their 30, 50 and 100 μm maps all have an elongated appearance similar to those of model 2. Furthermore, Gatley et al. noted that the position of the emission peaks in their maps varied with wavelength, and found that the 50 - 100 μm colour temperature decreased markedly towards the south-west whereas the optical depth at 50 μm was found to increase. Again, there is a qualitative similarity between the results of model 2 and the observations and deductions of Gatley et al., who concluded that M17SW is predominantly heated by external stars.

Although a number of approximations have been made in the models presented here, I would tentatively conclude that infrared observations of molecular clouds

do not necessarily imply the existence of internal heat
sources, and in the case of M17 it is highly likely that the
molecular cloud is heated externally.

References

Beckman, J. E., White, G. J., Cronin, N. J., and Chown, M. P.,
         1981, submitted to M.N.R.A.S.
Chown, M. P., Beckman, J. E., White, G. J., and Cronin, N. J.,
         This volume.
Gatley, I., Becklin, E. E., Sellgren, K., and Werner, M. W.,
         1979, Ap. J., 233, 575.
Natta, A., Palla, F., Panagia, N., and Preitte-Martinez, A.,
         1981, Astron. Astrophys., 99, 289.

DUST IN BOK GLOBULES

Iwan P. Williams
Applied Maths Dept. Queen Mary College, London E1 4NS
H. Bhatt
Physical Research Lab. Ahmedabad 380-009, India.

Abstract. The various forces which can operate on a
grain either in the vicinity or within a Bok globule are
reviewed and the consequences of various of these forces
being important discussed. A number of globules where
these forces may be in evidence are cited.

1. INTRODUCTION

Though most of the forces we shall discuss later operate,
to a larger or lesser extent, in all space we are in the main here
concerned with the interstellar medium in the vicinity of Bok globules
and within these globules. The existence of these quasi-spherical,
relatively opague clouds was pointed out over thirty years ago by Bok
and Reilly (1947). The fact that these condensations are gaseous
condensations where the high extinction is caused by dust grains is
generally accepted without question. What we shall discuss in this
paper is the possibility that, because certain forces are effective,
some relative motion of dust and gas has occurred which may lead to
an anomolous dust density distribution within some of these globules.

Of course, in order to investigate any particular globule
one should if possible insert as much structure to the gas within the
globule as observations [of the gas, not inferred from dust observations
will allow. Here we wish to discuss the types of situations which can
arise and will content ourselves with a simple model of a globule where
we take it to be spherical and stationary and where the gas density is
either constant or shows a simple power law variation. Grains will
also be regarded as spheres with a constant density. We shall ignore
magnetic fields for two reasons. First, it seems unlikely that at the

prevailing temperatures any substantial numbers of ionized matter will
be present. Secondly, even if fields are present, their main effect
is to cause charged particles to move along field lines. In general
this has the effect of reducing the efficiency of any driving force
by the cosine of the angle between the magnetic field and the force
[see for example Simons and Williams, 1979]

2. FORCES ON A GRAIN

As the forces are unrelated each will be discussed in turn
in sub paragraphs.

2.1 Gravity.

Gravity is an obvious force which exists within a globule.
Its direction is towards the centre of the globule and at distance R
away from the centre its magnitude is

$$\frac{Gm_g M_R}{R^2} ,$$

where G is the universal gravitational constant, M_g is the mass of
a grain and M_R is the mass within the radius R. With a power law for
the density of the form $\rho = \rho_0 \frac{R_0}{R}^n$, where ρ_0 and R_0 are constants,
then

$$M_R = \int_0^R 4\pi r^2 \rho dr = \frac{4\pi\rho_0 R_0^n R^{3-n}}{3-n}$$

It may also be convenient to express the grain mass as

$$M_g = \frac{4\pi a^3 \sigma}{3}$$

where a is the grain radius and σ its density. In this event

$$F_{Grav} = \frac{16\pi^2 \rho_0 \sigma a^3 R_0^n R^{1-n}}{3(3-n)}$$

2.2 Viscous Drag.

Whenever a body experiences a force which causes it to
move relative to an ambient gas, a viscous drag is experienced by the
body. In our case, the dust grain is always much smaller than the
mean free path within the gas and so the drag derived by Baines et al.
(1965) is valid. They give

$$F_{Drag} = \sqrt{\pi} \rho a^2 u^2 \left\{ \left(\frac{1}{S} + \frac{1}{2S^3}\right) e^{-S^2} + \left(1 + \frac{1}{S^2} - \frac{1}{4S^4}\right) \sqrt{\pi} \; \mathrm{erf}(S) \right\}$$

where S is a molecular speed ratio defined by $2u/W\sqrt{\pi}$, u being the
speed of the body relative to the gas and W the mean thermal velocity
in the gas. This is clearly a very unwieldy expression to use and two
approximations are of more use, namely S $\ll$ 1 (subsonic) and S $\gg$ 1
(supersonic). Performing the expansions, we have,

$$\text{subsonic, } F_{Drag} = \sqrt{\pi}\rho a^2 u^2 \left(\frac{8}{3S}\right) = \frac{4\pi\rho a^2 Wu}{3},$$

$$\text{supersonic } F_{Drag} = \pi\rho a^2 u^2.$$

2.3 Radiation Pressure.

Radiation Pressure has been recognized for some time as a
driving mechanism capable of generating high grain velocities. For
example Wickramasinghe (1972) claimed that velocities as high as
3×10^8 CM/S could be reached by grains in the neighbourhood of very
luminous stars. The generation of high velocity grains by radiation
pressure is also discussed by Tarafdar and Wickramasinghe (1976)
Simons and Williams (1976), Simons and Williams (1979). The force due
to radiation is given by

$$F_{Rad} = \pi a^2 QU\xi,$$

where U is the energy density of the radiation field, ξ a measure of
the anisotropy in the field and Q is the efficiency factor, which in
general is a complex function of grain radius composition and frequency
distribution of the radiation field. For a simple understanding of the
situation, it is sufficient to assume that Q $\sim$ 1 for a>10^{-1} μm and
Q $\sim$ 10^5a for smaller values of the grain radius.

It is obvious from the form of the expression that if
no anisotropy exists (ξ = o) no force exists. In practice, there are
two ways in which an anisotropy in the radiation field might arise.
First due to the nearby presence of a source of radiation, and
within the context of the present application, this is likely to be
because of the existence of an O or B star. Second, because the all
prevailing radiation throughout the galaxy (and in particular the
higher frequencies) has been blocked off from a particular direction,
leading to a resulting force towards the source of obscuration. Bok
globules obviously provide such a source of obscuration, its efficiency

as an obscurer dropping off as the inverse of the distance squared.

3. MOTION ARISING FROM THE FORCES.

The viscous drag discussed in 2.2 will of course cause motion of grains if a flow of gas exists, such as occurs during the interaction of the solar wind and a comet causing the dust and gas to flow outwards. Within the astronomical context of interest the viscous drag tends to oppose the motion of grains being set up by either of the other two forces. We again discuss the two cases in turn.

3.1 Drag opposing gravity.

This is a physical situation which has been studied in some depth within the context of planetary cosmogony from the situation in a spherical cloud by McCrea and Williams (1965) through a flattened cloud by Schatzmann (1971) to the situation in turbulent clouds by Simons et al. (1978).

An insight into the outcome can be obtained by assuming that the grain will reach its terminal velocity in a short time (which can be shown to be a correct assumption, see for example Williams and Crampin 1971) and thus equating the accelerating force given in 2.1 to the drag given in 2.2   yields an infall grain velocity (assumed subsonic, which turns out to be the case) of

$$u \; = \; \frac{4\pi\sigma aR}{(3-n)W} \quad ,$$

and thus a time to fall to the centre of

$$T \; = \; \frac{(3-n)W}{4\pi\sigma a} \; \sim \; \frac{500}{a} \quad \text{years}$$

With a normal grain size of 2×10^{-5} cm, all the grains initially present in a globule will be close to the centre in a period of 2.5×10^7 years. If grains are, or become, larger than normal then the time for settling decreases.

3.2 Drag opposing radiation pressure.

In this situation it is not so clear that grain motion will be subsonic. In their investigation, Simons and Williams (1976) showed that motion was likely to be supersonic in the neighbourhood of O B stars but subsonic due to the mean galactic radiation. The likely velocities in the two cases are again obtained by balancing the radiation force from 2.3 against the appropriate drag force from 2.2 .

This yields either

$$u = \frac{3QU\xi}{4W\rho} \quad \text{(subsonic)}$$

or

$$u = \left(\frac{QU\xi}{\rho}\right)^{\frac{1}{2}} \quad \text{(supersonic)}$$

and typical values range from 10^4CM/S due to the mean galactic radiation to 10^6CM/S or upwards resulting from the proximity of an O or B star. Since the lifetime of an O B star is short, both sources of driving grains will act over distances of the order of 10 pc so that an equivalent grain mass to that from a cloud with a mass in excess of $50M_\odot$ is available.

4.   SITUATIONS OF INTEREST

In addition to considering settling of grains and accretion from the surrounding medium, it is important to consider whether the globule has been in existence for a sufficiently long period for either, or both mechanisms to apply. We shall consider the possibilities in turn.

4.1   Young globule.

If the globule age is considerably less than 2×10^7y then grain settling will not occur to an extent which will be noticeable to an external observer. In this event, there are two possibilities to consider. Either there are O B stars nearby or there are not. If there are not, the amount of grains driven in from outside are likely to be small and so we would expect a globule with no abnormalities in the dust distribution and the dust to gas ratio.

If strong sources of radiation are present external to the globule, then we may expect an abnormally high dust concentration in the outer regions of a globule on the side lying nearest the sources. In fact the pressure from the grains being pushed in may also deform the globule. It is the extinction that has been measured in general. As this depends on both the density and the length of the sight line, increasing the density in the outer regions tends to flatten the extinction profile. In addition of course the value of the extinction is increased at all points. Globules B68 and Ori-2 conform with this picture.

4.2   Old Globules.

In globules whose ages exceed about 2×10^7y substantial grain settling will have occurred. The extinction profile of such a

globule depends on whether or not the grain population in the outer
regions has been replenished by the introduction of grains driven into
the globule from the external medium. If this has not occurred, then
the globule will be devoid of grains in its outer regions while a
dense core will exist. Globules B227 and B335 could be thought of as
having such an appearance.

On the other hand, if grains are being driven into
the globule, then a calculation is necessary to determine the expected
extinction profile. This calculation has been carried out by
Williams and Bhatt (1982) and this shows that the density of grains
within a globule will vary approximately as the inverse cube power of
the radius. The implied extinction profile was found to be in very
good agreement with the observed (from star count) extinction profiles
for globule B361 (Schmidt, 1975) and globule 2 in the coalsack (Bok,
1977). The only other explanation offered for the extinction profile
in these two globules is by Kenyon and Starrfield (1979) who assume
a polytropic model for these globules with a slightly unusual polytropic
index but with a normal dust to gas ratio.

5. CONCLUSIONS AND DISCUSSION

Provided globules are not in a state of free fall collapse,
grains will settle towards the centre in the manner indicated. Rotation
and magnetic field alter the detailed picture and the mathematical
calculations but do not alter the general conclusions. Grain growth,
either by coagulation or condensation, both of which are possibilities,
will decrease the grain settling time. However, turbulence within the
globule will tend to prevent settling and indeed if turbulence with a
typical velocity of 1km/s were to be present, grain radii would have
to be of the order of 100 times greater than normal before the settling
velocity exceeded the random turbulent velocity. It must be remembered
however that anomalously large grain radius would only be detectable
in the infra red.

Similarly, grains being driven by radiation pressure will
be present and so it seems very likely that most of the situations
described above will come about in some globule somewhere. Of course
if the grains are driven by strong radiation such as from an O B star,
they may sweep some of the interstellar medium with them, thus cause
pressure to occur at the region of impact with the globule and so

cause considerable deformation to the shape of the globule. All these
modifications will be investigated in detail at a later stage.

It will also be instructive to carry out detailed
observations in the infra red, and using molecular lines, of a small
number of potentially interesting globules to determine column density
profiles,average grain sizes and dust to gas ratios.

Acknowledgement

This work became possible as a result of a British Council
grant to allow collaboration between Queen Mary College and the
Physical Research Laboratory.

References

Baines, M.J., Williams, I.P., & Asebiomo, A.S. (1965) Mon. Not. R. astr. Soc., 130, 63.

Bok, B.J. (1977) Pub. A.S.P., 89, 597.

Bok, B.J., & Reilly, E.F. (1947) Ap. J. 105, 255

Kenyon, S., & Starrfield, S. (1979) Pub. A.S.P., 91, 271.

McCrea, W.H., & Williams, I.P. (1965) Proc. Roy. Soc., A287, 143

Schatzmann, E. (1971) Physics of the Solar System, Goddard Inst. X650-71-3.

Schmidt, E.G. (1975) Mon. Not. R. astr. Soc., 172, 401.

Simons, S., & Williams, I.P. (1976) Astroph. Sp. Sci., 43 239.

Simons, S., & Williams, I.P. (1979) Astroph. Sp. Sci., 61 411.

Simons, S., Simpson, I.C., & Williams, I.P. (1978) Moon and Planets, 19, 399.

Tarafdar, S.P., & Wickramasinghe, N.C. (1976) Astroph. Sp. Sci., 39, 19.

Wickramasinghe, N.C. (1972) Mon. Not. R. astr. Soc., 159, 269.

Williams, I.P., & Bhatt, H. (1982) Mon. Not. R. astr. Soc. 199, 465

Williams, I.P., & Crampin, D.J. (1971) Mon. Not. R. astr. Soc., 152, 261

SECTION II

Spectroscopic observations of molecular sources

AIRBORNE FAR-INFRARED AND SUBMILLIMETER SPECTROSCOPY

Dan M. Watson
Department of Physics, University of California,
Berkeley, California 94720, U.S.A.

Abstract. This paper is a review of airborne spectroscopic observations in the wavelength range 18 μm - 800 μm, and the instruments on the NASA Kuiper Airborne Observatory with which they were carried out. Recent measurements of spectral lines of ions, atoms and molecules are presented and discussed.

Airborne Astronomy

Except for a few very poor windows, the earth's atmosphere is opaque over the range 14 μm $< \lambda < 1$ mm. Unfortunately, many interesting astronomical objects emit most of their radiation in this range, and the lowest-lying transitions of many astrophysically significant molecules, atoms and ions also occur at these wavelengths -- in fact, it is the presence in our atmosphere of one astrophysically interesting molecule, H_2O, that is responsible for the opacity. Water, however, is mostly confined to the lower layers of the atmosphere. For altitudes just a thousand meters or so above the tropopause, the atmospheric optical depth is about three orders of magnitude lower than at sea level.

It was to provide a sizeable telescope platform at such altitudes that the U.S. National Aeronautics and Space Administration (NASA) built the Gerard P. Kuiper Airborne Observatory (KAO). The KAO, shown in Figure 1, is a specially-modified Lockheed C-141 equipped with a gyrostabilized, computer-guided 91.4 cm telescope of optical quality figure. Inside the pressurized cabin (Figure 2), astronomers can work alongside their instruments in a "shirt-sleeve" environment. The KAO joins the NASA LearJet observatory and a handful of balloon-borne telescopes in the ranks of high-altitude infrared observatories, and in the past few years has become the leading instrument in far-infrared astronomy.

Because of this versatile observatory and recent advances in the technology of receivers and detectors, it is now possible to exploit the far-infrared and submillimeter spectra of ions, atoms, and molecules in the study of the interstellar medium. This paper is a review of instruments and observations on the KAO, and is written in the wake of the first very rapid burst of activity in what promises to be a field of prime importance in astrophysics. After a survey of the spectrometers currently in use, sections are included in which results on ions, neutral atoms, and molecules are discussed. The paper concludes with a bibliography of airborne far-infrared and submillimeter spectroscopic results.

Figure 1: NASA's Gerard P. Kuiper Airborne Observatory (KAO). The observatory's 91.4 cm telescope looks out through the open port visible on the airplane's side, forward of the wing (NASA photograph).

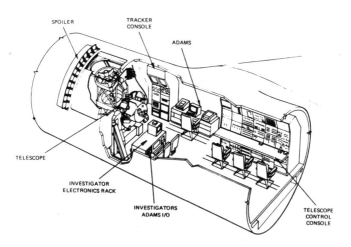

Figure 2: Interior of the KAO. The ADAMS (Airborne Data Acquisition and Management System) is the computer that controls the telescope; it is also used to acquire, store and process data in flight.

Spectroscopic Instrumentation on the KAO

There are no observatory-supported instruments for the KAO at the present time, but several spectrometers have been built by university and industrial research groups, and the KAO's first "facility" far-infrared spectrometer is now under development at NASA Ames Research Center. The devices already in use represent efforts to extend incoherent detection and optical tecniques into the far-infrared as well as the application of coherent detection and other common radiofrequency techniques to shorter submillimeter wavelengths. A list of these instruments and some of their parameters appears in Table 1.

The recent successes of incoherent receivers are due in large part to advances in the technology of far-infrared detectors. Development programs in several places, particularly U.C. Berkeley and Cornell University, have produced extrinsic germanium photoconductors that are several orders of magnitude more sensitive at low flux levels than bolometers. The most sensitive photoconductors are Ge:Ga, which responds at wavelengths shortward of about 115 μm (Haller *et al.* 1979), and Ge:Sb, which goes out to 135 μm (Watson *et al.* 1980a). Longer-wavelength response may be obtained from p-type germanium photoconductors by applying mechanical stress (Haller *et al.* 1979). In stressed Ge:Ga, for instance, response out to 230 μm has been obtained (R.W. Russell, personal communication).

In the design of incoherent receivers, two considerations dictate the type of spectrometer to use. First, far-infrared detectors are sufficiently advanced that photon noise in

Table 1 Spectrometers on the KAO			
Wavelength range (μm)	Resolution ($\lambda/\Delta\lambda$)	Description of instrument	Institution, references
18-40	800	Cooled grating, linear array of Ge:Ga or Si:Sb photoconductors.	Cornell U. (McCarthy, Forrest & Houck 1979)
40-200	100-1000	Cooled grating and Ge:Ga photoconductor, "lamellar" interferometer.	Cornell U. (Houck and Ward 1979; Harwit *et al.* 1981)
18-200	5000	Michelson interferometer, bolometer.	ESTEC/Meudon (Baluteau *et al.* 1976)
40-130	4000	Tandem Fabry-Perot interferometers, Ge:Ga or Ge:Sb photoconductor.	U.C. Berkeley (Storey, Watson, & Townes 1980)
18-115	10000-20000	Cooled grating, linear array of Ge:Ga photoconductors.	NASA Ames (E.F. Erickson, personal communication)
300-1000	5×10^5	InSb hot-electron bolometer mixer, multiplied klystron LO.	CalTech/Bell Labs (Phillips & Jefferts 1973)
300-1000	5×10^5	InSb hot-electron bolometer mixer, carcinotron LO.	ESTEC (van Vliet *et al.* 1981)

the thermal emission of warm optical elements in the detector's field of view dominates the system noise (The term "background-limited" is often used to describe this situation). Next, thermal emission from dust in molecular clouds is very strong in this part of the spectrum, and faint molecular lines would be hard to detect against this bright continuum. Narrow-band instruments, such as diffraction gratings and Fabry-Perot interferometers, satisfy the conditions of reduced background and continuum, and are generally favored. In particular, under background-limited conditions it is a handicap to expose the detector to a large bandwidth and precede it with a multiplexing interferometer, contrary to what is advantageous in the near-infrared.

A diagram of the U.C. Berkeley tandem Fabry-Perot spectrometer (Storey, Watson & Townes 1980) is shown in Figure 3 as an example of a low-background, high-resolution incoherent receiver. A single liquid helium-cooled Ge:Ga or Ge:Sb photoconductor, mounted in an integrating cavity, is the detector. The detector is followed by a transimpedance preamplifier, the first stage of which is a liquid nitrogen-cooled JFET. Preceding the detector, in turn, are: a wheel carrying four fixed-wavelength Fabry-Perot interferometers (FPI's), each set in first order at a wavelength of interest, any of which can be rotated into the beam for

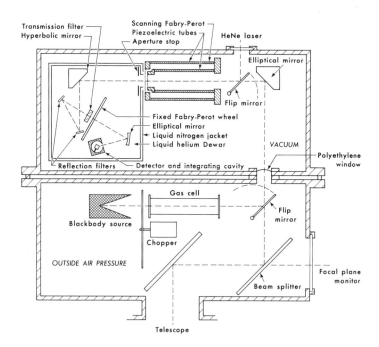

Figure 3: Schematic diagram of the U.C. Berkeley tandem Fabry-Perot spectrometer (Storey, Watson & Townes 1980).

observations; a series of antireflection-coated salt crystals and diamond-dust-bearing pellicles, which serve as low-pass filters and reject the higher orders of the fixed-wavelength FPI's; and another FPI tuned so that one of its higher orders (usually 20-50) lies within the passband of the fixed-wavelength FPI; the high-order FPI is scanned piezoelectrically within this passband. The FPI mirrors are made of free-standing metal mesh, with high reflectivity and very low absorption, giving the FPI's finesses up to about 120 and peak transmissions in the range 50-80%. The fixed-wavelength FPI's and lowpass filters are cooled with liquid helium, and the scanning FPI is placed immediately outside the liquid helium-cooled section of the system, with cooled shrouds nearly touching the mirrors. Thus the background radiation incident on the detector is reduced to that transmitted by the single selected order of the scanning FPI. In its present configuration, the spectrometer is capable of a resolving power of a part in 4000, and has an NEP of 5×10^{-14} W Hz$^{-1/2}$, including all spectrometer, telescope, and atmospheric losses.

In the submillimeter region, both of the receivers in use on the KAO are heterodyne spectrometers based on InSb hot-electron bolometer mixers. Figure 4 is a schematic of one of these devices, developed at Bell Telephone Laboratories by Phillips & Jefferts (1973). The mixers have bandwidths of about 1 MHz; they are used with tunable local oscillators which are scanned in frequency to move the detector response across the profile of the spectral line being observed. System noise temperatures of 300-1000 K are typical of the submillimeter performance of these receivers.

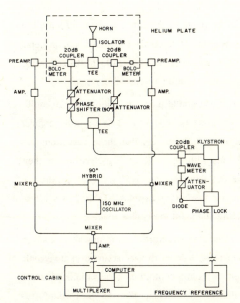

Figure 4: Block diagram of the InSb hot-electron bolometer heterodyne receiver developed by Phillips and Jefferts (1973).

In principle, coherent spectral-line receivers have a substantial advantage over incoherent receivers in the far-infrared and submillimeter, because they lend themselves more easily to very high spectral resolution and correspondingly smaller thermal background noise. (It is interesting to note that continuum emission from warm molecular clouds may itself provide significant background in some cases; thus for molecular line observations of warm dense clouds a heterodyne system on the KAO may be competitive with a spectrometer on a cryogenically-cooled telescope). It would be desirable to have wide-bandwidth heterodyne systems throughout this wavelength range. An example of this sort of system is the submillimeter receiver of Goldsmith *et al.* (1981), which is based on a Schottky diode mixer and molecular laser local oscillator, and has been used for observations of the $J = 6 \rightarrow 5$ line of CO at 434 μm. Perhaps at the shorter wavelengths ($< 200 \mu$m) extrinsic germanium photoconductors will perform well as mixers, and heterodyne systems similar to those operating in the 10 μm region can be developed. More discussion of the nature of future high-frequency receivers can be found in the contributions of Encrenaz and de Graauw in this volume.

While we await the development of coherent systems in the far-infrared, there is still much science to be done with incoherent spectrometers with spectral resolutions of a part in $1-3 \times 10^4$, and detector arrays will also increase the power of these systems.

Lines Arising From Ionized Regions

The strongest lines in the infrared spectrum of the interstellar medium are the ionic fine structure lines of H II regions. The properties of these lines are discussed in detail elsewhere (cf. Watson & Storey 1980, Simpson 1975, and references therein). In the following, we will survey the specifics of the far-infrared lines that require airborne observation.

Fine structure lines are produced by magnetic dipole transitions within multiplets of given spin and orbital angular momentum whose states of different total angular momentum are split by the spin-orbit interaction. In ionized nebulae these transitions are excited by ion-electron collisions. Atoms and ions with 1, 2, 4, or 5 p-electrons in their valence shells have ^2P or ^3P ground states (like the abundant ions N^{++} and O^{++}, pictured in Figure 5), and hence have astronomically important fine structure lines. Table 2 is a list of far-infrared fine structure lines for some of the more abundant interstellar species.

An important property of far-infrared fine structure lines is that ionic column densities can be obtained directly from their intensities, leading to very accurate relative ionic and elemental abundances (and, from comparison with radio continuum observations, absolute abundance.) The long wavelengths and small radiation rates indicate that the lines would usually be unextinguished and optically thin; therefore the power per unit area emitted in the tran-

Table 2 Fine Structure Lines Arising In Ionized Regions						
Transition	Species	λ (μm)	A (s^{-1})	Excitation Potential (eV)	Ionization Potential (eV)	Seen?
$2p$: $^2P_{3/2} \rightarrow {}^2P_{1/2}$	N^{++}	57.30	4.77×10^{-5}	29.601	47.448	yes
	O^{+++}	25.91	5.18×10^{-4}	54.934	77.413	yes
$3p$: $^2P_{3/2} \rightarrow {}^2P_{1/2}$	S^{+++}	10.52	7.70×10^{-3}	34.83	47.30	yes
$2p^2$: $^3P_1 \rightarrow {}^3P_0$	N$^+$	203.9	2.13×10^{-6}	14.534	29.601	no
$^3P_2 \rightarrow {}^3P_1$		121.7	7.48×10^{-6}			no
	O^{++}	88.356	2.62×10^{-5}	35.117	54.934	yes
		51.815	9.75×10^{-5}			yes
$3p^2$: $^3P_1 \rightarrow {}^3P_0$	S^{++}	33.443	4.72×10^{-4}	23.33	34.83	yes
$^3P_2 \rightarrow {}^3P_1$		18.713	2.07×10^{-3}			yes
$2p^4$: $^3P_1 \rightarrow {}^3P_2$	Ne^{++}	15.55	5.99×10^{-3}	40.962	63.45	no
$^3P_0 \rightarrow {}^3P_1$		36.04	1.15×10^{-3}			no
$3p^4$: $^3P_1 \rightarrow {}^3P_2$	Ar^{++}	8.99	3.08×10^{-2}	27.629	40.74	yes
$^3P_0 \rightarrow {}^3P_1$		21.83	5.19×10^{-3}			no
$2p^5$: $^2P_{1/2} \rightarrow {}^2P_{3/2}$	Ne$^+$	12.81	8.59×10^{-3}	21.564	40.962	yes
$3p^5$: $^2P_{1/2} \rightarrow {}^2P_{3/2}$	Ar$^+$	6.99	5.26×10^{-2}	15.759	27.629	yes

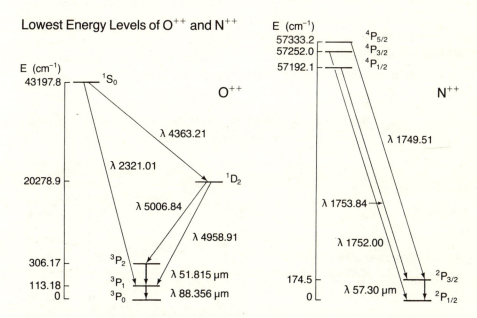

Figure 5: Energy level diagrams for the lowest-lying states of O^{++} and N^{++} (not drawn to scale). The ground-state fine structure lines of each ion are shown, as well as the familiar visible lines of O^{++}, $\lambda\lambda$ 5007, 4959 ("nebular") and λ 4363 ("auroral"). N^{++} has no forbidden lines in the visible spectrum.

sition from state k to state j can be written as:

$$I_{kj} = \frac{hc}{4\pi\lambda} A_{kj} N_k \Delta\Omega \quad , \tag{1}$$

where A_{kj} is the A-coefficient, $\Delta\Omega$ is the beam area, and N_k is the column density in state k of the ion, averaged over the beam. The partition of ions over the fine structure levels can be determined from the statistical equilibrium of collisional and radiative excitation and de-excitation:

$$N_k \left\{ \sum_{j<k} A_{kj} + n_e \sum_j \gamma_{kj} \right\} = \sum_{j>k} N_j A_{jk} + \sum_j N_j n_e \gamma_{jk} \quad , \tag{2}$$

where n_e is the density of electrons and γ_{kj} is the electron collisional rate coefficient for the transition $k \to j$. The rate coefficients can be computed theoretically to an accuracy of 10-20%. The population of the upper multiplets can nearly always be ignored, leaving a system of two or three equations which can be solved for the fine structure level populations at a given electron density. Since the level spacings are much smaller than kT for H II regions (which typically have $T \approx 8000$ K), the results depend very weakly on temperature. Figure 6 shows the solutions of Equation 2 for the ions O^{++} and N^{++}. Note that the level populations approach their thermal equilibrium values at quite low densities; the total ionic column density can often be obtained from the line intensities by assuming thermal equilibrium. For the cases where the levels would not be thermalized, it is possible to use the intensity ratio of the two fine structure lines belonging to an ion with a triplet ground state to determine the electron density and thus derive the level populations. The result, which can be derived by using the solutions of Equation 2 in Equation 1, is:

$$n_e = \frac{\gamma_{01}(A_{21} + A_{20}) + \gamma_{02}[A_{21} + (R/a)A_{10}]}{\gamma_{01}[(R/a)\gamma_{12} - \gamma_{21} - \gamma_{20}] + \gamma_{02}[(R/a)(\gamma_{12} + \gamma_{10}) - \gamma_{21}]} \quad , \tag{3}$$

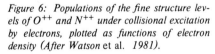

Figure 6: Populations of the fine structure levels of O^{++} and N^{++} under collisional excitation by electrons, plotted as functions of electron density (After Watson et al. 1981).

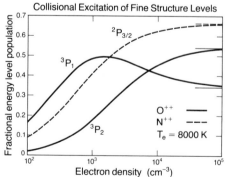

where the levels have been labelled 0, 1, 2 in order of increasing energy, $R = I_{10}/I_{21}$, and $a = \lambda_{21}A_{10}/\lambda_{10}A_{21}$; this result is independent of ionic abundance and source geometry. The two fine structure lines of O^{++} at 88.4 μm and 51.8 μm are often used in this manner to derive electron densities.

Far-infrared fine structure lines have a strong influence on the overall thermal equilibrium of H II regions through the cooling they provide. The lines of O^{++} by themselves provide about one third of the total cooling rate in typical H II regions (cf. Osterbrock 1974). There is, therefore, much to be learned about the energetics and dynamics of ionized regions, and their interaction with ambient material, from observations of far-infrared lines.

Because the fine structure lines are relatively strong and spectrometers have improved dramatically (and are still improving rapidly), their detection in galactic sources has become relatively easy. Figure 7 shows some typical observations of the two [O III] lines and the [N III] line; the integration times for these measurements were just a few minutes. It is now possible to observe fine structure lines in H II regions over most of the galaxy. This opens

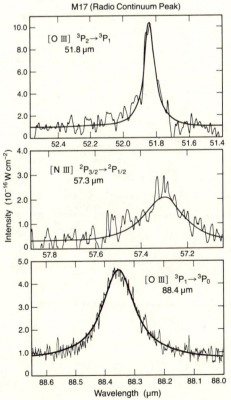

Figure 7: Fine structure line emission from M17, measured in a 1' beam. The lines are unresolved; the curves drawn through the data are lorentzians, approximating the instrumental transmission function (After Storey et al. 1979, Watson et al. 1981).

up the intriguing prospect of measuring possible variations of elemental abundance in the galaxy, as is already being attempted at visible and mid-infrared wavelengths (eg. Lester *et al.* 1981 and references therein). Because of the very low interstellar extinction at far-infrared wavelengths and the relatively straightforward interpretation of fine structure line fluxes, these lines represent the best method available for measuring the galactic distribution of the abundances of the elements N, O, and S.

Lines Arising in Neutral Atomic Regions

As is the case in ionized regions, the strongest infrared lines from neutral atomic regions are fine structure lines (A list of these appears in Table 3). These lines are collisionally excited and are an important cooling mechanism in atomic regions such as ionization fronts, diffuse shocked gas, and diffuse clouds (H I regions). In principle, density and column density can be derived from these line fluxes, as is done with the lines from ionized regions, although it is more difficult than in the case of H II regions for three reasons: the relative importances of the available collision partners (e, p, H, H_2) are not obvious; the line fluxes are more sensitive to temperature in most cases; and the collision rates have not yet been computed as accurately. In spite of these difficulties, observations of neutral species have produced some of the most interesting results yet obtained in this wavelength region, as the following three examples illustrate.

Table 3
Fine Structure Lines Arising in Neutral Regions

Transition	Species	λ (μm)	A (s^{-1})	Excitation Potential (eV)	Ionization Potential (eV)	Seen?
$2p$: $^2P_{3/2} \rightarrow {}^2P_{1/2}$	C$^+$	157.4	2.36×10^{-6}	11.260	24.383	yes
$3p$: $^2P_{3/2} \rightarrow {}^2P_{1/2}$	Si$^+$	34.8	2.13×10^{-4}	8.151	16.345	no
$2p^2$: $^3P_1 \rightarrow {}^3P_0$ $^3P_2 \rightarrow {}^3P_1$	C^0	609.133 370.414	7.93×10^{-8} 2.68×10^{-7}	...	11.260	yes no
$3p^2$: $^3P_1 \rightarrow {}^3P_0$ $^3P_2 \rightarrow {}^3P_1$	Si^0	129.68 68.474	8.25×10^{-6} 4.20×10^{-5}	...	8.151	no no
$2p^4$: $^3P_1 \rightarrow {}^3P_2$ $^3P_0 \rightarrow {}^3P_1$	O^0	63.170 145.526	8.95×10^{-5} 1.70×10^{-5}	...	13.618	yes yes
$3p^4$: $^3P_1 \rightarrow {}^3P_2$ $^3P_0 \rightarrow {}^3P_1$	S^0	25.246 56.322	1.40×10^{-3} 3.02×10^{-4}	...	10.360	no no

Neutral Oxygen

The most luminous [O I] source yet observed is the galactic center, Sgr A. The [O I] 63.2 μm line, first observed in Sgr A by Lester *et al.* (1981), revealed for the first time the presence of neutral gas within the central few parsecs of the galaxy, with velocity dispersions comparable to those of the ionized gas in the very center. Further observations (Genzel *et al.* 1982, in preparation) have produced a map of the [O I] emission (Figure 8) which shows the source to be extended over about 12 pc along the galactic plane and have an integrated brightness of about 10^5 solar luminosities in the line. There is a line center velocity gradient across the source which is symmetrical about the galactic center, indicating rotation in the same sense as the rest of the galaxy. The large linewidths ($\geqslant$ 200 km s^{-1}) and the net redshift of the source ($\approx$ +50 km s^{-1}) indicate large radial motions, and other noncircular motions, as well. In general the [O I] source resembles a rotating, expanding (or contracting) ring of neutral gas. This heretofore unknown component of the galaxy is thus reminiscent of the well-known expanding molecular features at galactocentric radii 180 pc and 4 kpc (cf. Oort 1977).

Ionized Carbon

The [C II] line, considered to be one of the most important cooling lines in the interstellar medium, has been observed in the directions of several H II regions and found to be distributed over areas much larger than the H II region itself in each case. In M 17, for instance, a strip map taken along a direction where the width of the H II region is 3' showed

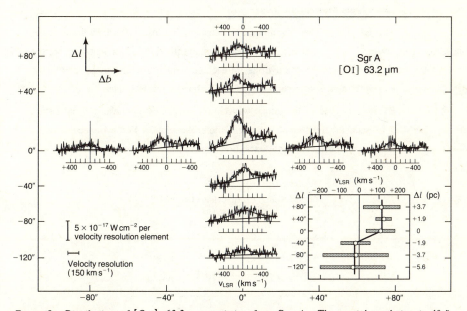

Figure 8: Distribution of [O I] 63.2 μm emission from Sgr A. The spatial resolution is 40 " (FWHM). A longitude-velocity diagram is shown in the inset; the horizontal bars indicate the linewidths (FWHM) at each position (Genzel et al. 1982, in preparation).

that the [C II] extended at least 15′ from the radio continuum peak; observations of C^+ recombination lines, however, show only a localized region at the edge of the H II region (Russell *et al.* 1981). It appears that M 17 is surrounded by an enormous neutral atomic envelope not previously known to exist; this may turn out to be a common phenomenon.

Neutral Carbon

Before the [C I] line was detected (by Phillips *et al.* 1980b), it was thought to be very weak, because neutral atomic carbon was presumed to exist only in thin rims on the edges of molecular clouds (within the ultraviolet penetration depth), and formed from CO dissociated by starlight. However, widespread [C I] emission has since been observed in many molecular clouds (cf. Phillips & Huggins 1981), and the lines are far too strong to be accounted for by the above picture. In fact, the atomic carbon column densities are found to be comparable in size to CO column densities in each case where enough data exist to make the comparison (see Table 4). It appears that the simple, static, stratified models of molecular clouds and their chemistry are inadequate. Suggestions have been made (Phillips & Huggins 1981) that molecular clouds may simply be too young ($\leqslant 10^6$ yrs) to have reached steady state and have their interiors be completely shielded, or that "rejuvenation" processes take place in molecular clouds, perhaps by a convective dredging of molecular matter to the edges where it can be photodissociated, or by intermittent production of ultraviolet radiation within the cloud by shocks or young stars.

Lines From Molecular Clouds

The far-infrared and submillimeter wavelength range is rich with molecular lines, especially those of light molecules and radicals. Among these lines, for instance, are the lowest pure rotational transitions of the hydrides of each element in the first three rows of the periodic table, along with their isotopes and ions; Table 5 shows a few examples.

Except for some of the longer wavelength ones, most of these molecular lines are not likely to be seen in emission in quiescent molecular clouds. For a molecule with a dipole moment of 1 debye, the collisional excitation rate for a transition of wavelength λ is

Table 4 CO and Neutral Carbon Column Densities*				
Source	τ_{CO}	τ_{C^0}	N_{CO} (cm^{-2})	N_{C^0} (cm^{-2})
OMC-1	80.	25.	5.0×10^{19}	3.5×10^{19}
NGC 2024	33.	4.	2.1×10^{18}	1.5×10^{18}
NGC 2264	25.	1.3	1.2×10^{18}	0.5×10^{18}

* From Phillips and Huggins (1981).

equal to the raulation rate at a density of roughly $n = (3 \times 10^9 \, \text{cm}^{-3}) \left[\dfrac{100 \mu \text{m}}{\lambda} \right]^3$; that is, collisionally-excited emission in a cool cloud requires a density so high that most of such regions would be optically thick in dust and the emission would not be observed. Absorption lines from such regions would also not be observed, since the dust and gas would be in thermal equilibrium at such high densities. However, absorption lines due to molecules in the cooler, outer parts of the cloud could be seen, as is the case in the Λ-doubling lines of some of these light molecules at longer wavelengths (eg. OH, CH).

The far-infrared pure rotational lines have two observational advantages over the Λ-doubling lines. First, the sources of far-infrared and submillimeter continuum radiation (thermal emission from dust) are highly correlated in position with the molecules, which is not necessarily the case for the radio continuum sources. Second, the optical depth of the rotational lines is much greater than that of the radiofrequency Λ-doubling lines. For the ground rotational state, the Λ-doubling levels will usually be in thermal equilibrium because of their

Table 5 Rotational Lines Of Some Of The More Abundant Hydrides	
Molecule, lowest transition	λ (μm)
HCl $J = 1 \rightarrow 0$	479.0
SH $^2\Pi_{3/2}$ $J = \frac{5}{2} \rightarrow \frac{3}{2}$	218.8
H$_2$S $J_{K^+K^-} = 1_{01} \rightarrow 0_{00}$	421.0
SiH $^2\Pi_{1/2}$ $J = \frac{3}{2} \rightarrow \frac{1}{2}$	439.7
AlH $J = 1 \rightarrow 0$	805.8
MgH $^2\Sigma^+$ $J = \frac{1}{2}$, $N = 1 \rightarrow 0$	871.1
NaH $J = 1 \rightarrow 0$	1002.8
OH $^2\Pi_{3/2}$ $J = \frac{5}{2} \rightarrow \frac{3}{2}$	119.4
H$_2$O $J_{K^+K^-} = 1_{01} \rightarrow 0_{00}$	429.3
NH $^3\Sigma^-$ $N = 1 \rightarrow 0$, $J = 2 \rightarrow 1$	307.6
NH$_3$ $J_K = 1_0 \rightarrow 0_0$	523.7
CH $^2\Pi_{3/2}$, $J = \frac{3}{2} \rightarrow {}^2\Pi_{1/2}$, $J = \frac{1}{2}$	561.4
HD $J = 1 \rightarrow 0$	112.1
H$_2$ $J = 2 \rightarrow 0$	28.2

close spacing, while the next higher rotational state will have a much lower population than the ground state, as we saw above; under these conditions it can be shown that the ratio of optical depths of rotational and Λ-doubling lines is

$$\frac{\tau_{rot.}}{\tau_\Lambda} \approx \frac{kT}{h\nu_\Lambda}$$

In the case of OH, for instance, this ratio is at least 40; this is demonstrated by the OH spectrum in Figure 9 (Watson *et al.* 1982, in preparation), which shows very optically thick OH rotational lines from a region in the Sgr B2 molecular cloud where the OH Λ-doubling optical depth is only about 0.1. The consequence of this is that smaller amounts of molecules may be detected in the far-infrared and submillimeter range than in the radio. Because the molecules, radicals and molecular ions that can be observed in this range include many of the building blocks of the larger molecules observed at lower frequencies, molecular observations in the far-infrared and submillimeter will be of great significance in the study of interstellar chemistry.

Far-infrared molecular emission lines can be produced by energetic phenomena within molecular clouds. There are several examples known of regions in molecular clouds which have been compressed and heated by shock waves; in most cases the shocks are thought

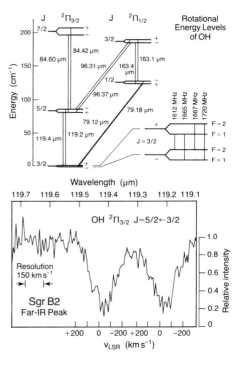

Figure 9: (a) The lowest rotational energy levels of the hydroxyl radical, OH. Each level is split by Λ-doubling and hyperfine structure, as shown for the $^2\Pi_{3/2}$ J= 3/2 state. Also shown are the familiar radio-frequency lines in this state. (b) OH absorption lines at 119.2 and 119.4 μm in Sgr B2. The beam size was 40 " (FWHM). Hyperfine structure in these lines is not visible because of instrumental resolution and motions in the source. Accounting for the resolution, the opacity in the lines is essentially 100%. The lines are centered at $v_{LSR} = -80$ km s^{-1}, and are probably due primarily to absorption by the so-called "molecular ring" which lies at a galactrocentric radius of 180 pc (Watson et al. 1982, in preparation).

to be driven into the ambient molecular matter by mass outflow from a newly-formed massive star (cf. Genzel & Downes 1981, and references therein). The post-shock matter can be quite hot, up to 2000-3000 K, with densities higher than ambient as well, and far-infrared/submillimeter emission lines can be excited, both in the transitions of lighter molecules like those in Table 5 and in the higher transitions of heavier molecules, notably CO. The archetypal shocked region is the Kleinmann-Low region in Orion, where, in addition to near- and mid-infrared H_2 emission and high-velocity emission in microwave and millimeter-wave lines, many submillimeter and far-infrared molecular lines have been observed. Best represented is the CO molecule, for which the $J = 4{\rightarrow}3$ (Phillips *et al.* 1980a), $6{\rightarrow}5$ (Goldsmith *et al.* 1981), $17{\rightarrow}16$ (Harwit 1981, personal communication), $21{\rightarrow}20$, $22{\rightarrow}21$, $27{\rightarrow}26$, and $30{\rightarrow}29$ (Watson *et al.* 1980a; Storey *et al.* 1981b) lines have been observed, as well as the $J = 3{\rightarrow}2$, $2{\rightarrow}1$ and $1{\rightarrow}0$ lines. Emission from H_2O (Phillips *et al.* 1980a) and OH (Storey *et al.* 1981a) has also been reported. The rotational lines of CO, OH, and H_2O are very important in cooling the shocked gas (cf. Hollenbach & McKee 1979), and other far-infrared/submillimeter lines are potentially useful as probes of the unusual chemical conditions there. Being largely free from the problems of extinction and contamination by emission and absorption from gas along the line of sight which respectively affect the interpretation of near-/mid- infrared and millimeter observations, the far-infrared seems uniquely well-suited for the study of these regions of active star formation.

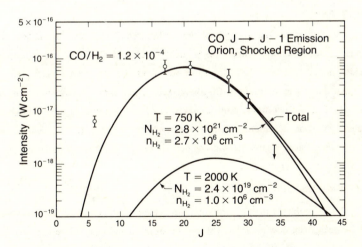

Figure 10: Distribution of CO line intensities in the shocked region of the Orion molecular cloud, together with parameters derived from the intensities. The points at $J = 30$, 27 and 21 come from Storey et al. *(1981), the one at $J = 17$ from Harwit (1981, personal communication), and the one at $J = 6$ from Goldsmith* et al. *(1981). The latter line shows the influence of cooler gas further behind the shock than the 750 K and 2000 K components.*

The far-infrared lines of CO, in particular, have proven very useful in the Orion problem. The cross-sections for collisional excitation of the CO rotational levels can be computed to high accuracy (≈ 30 %) even for rotational quantum numbers as high as $J = 60$ (Green & Thomas 1980; McKee et al. 1981). It turns out that the collisional transition rates and radiation rates are equal when $J \approx 25$ for the likely density and temperature conditions in the shocked gas, so the observations, involving J in the range 20-35, include lines that are far from thermal equilibrium as well as lines that are very close. Thus a comparison between observations and multilevel calculations of CO emission using accurate collision cross-sections allows a determination of H_2 density, temperature, and total CO column density in the shocked gas. Furthermore, comparison of the total CO and H_2 column densities (the latter derived from 2 μm and 12 μm H_2 line observations), gives the CO abundance. Storey et al. (1981b) used this procedure in a two-component model of the Orion shocked region, assuming pressure equilibrium between the two components and taking $T = 2000$ K for the hotter component, as indicated by the 2 μm H_2 observations. The results of an improved version of this calculation is presented in Figure 10; the major difference between the present result and the previous one is the inclusion of more extensive 12 μm H_2 line data (Beck et al. 1982) and an improved value for the extinction toward the H_2 source (Beckwith et al. 1981). The quality of the fit to the data is such that the H_2 density and CO/H_2 relative abundance are determined to within a factor of two of 1×10^6 cm^{-3} and 1.2×10^{-4} respectively. Most important is the density determination, the first such determination in the shocked region; the derived H_2 density is too small by about an order of magnitude for the observed H_2 and CO emission to be produced by a purely gas-dynamic shock.

· A very interesting sidelight of the far-infrared CO observations is the CO/H_2 relative abundance determination. Since the H_2 and far-infrared CO lines are optically thin, arise in shielded matter and arise from the same well-defined component of gas, the results constitute the first direct measurement of CO abundance. There is no evidence which indicates that significant H_2, CO, or dust dissociation has taken place with the passing of the shock wave, and we conclude that this CO abundance may also apply to the ambient molecular medium. There are many other cases where the well-defined conditions in shocked molecular gas can lead to direct abundance determinations.

Prospects

Many problems in star formation, interstellar chemistry, and the structure and evolution of our galaxy can now be addressed with far-infrared and submillimeter spectroscopy. There is still a vast amount of work that can be done with existing instruments, particularly since the southern hemisphere has yet to be explored. Current observations focus on abundances in H II regions, the dynamics of shocked matter in molecular clouds, and the interstellar abundances of ionized and neutral carbon.

There are still some orders of magnitude of sensitivity to be gained, though, through the use of heterodyne receivers on airborne observatories and through the eventual availability of cryogenically-cooled spaceborne telescopes like NASA's Shuttle Infrared Telescope Facility (SIRTF). These advances may permit far-infrared and submillimeter spectroscopic observations of extragalactic objects. In Berkeley, an effort is already under way to develop heterodyne receivers in the 50 μm - 800 μm region, and it is hoped that the next few years of work may produce such a system for the KAO.

Acknowledgments

The new observations with the U.C. Berkeley tandem Fabry-Perot spectrometer presented in this paper were carried out in collaboration with R.L. Genzel and C.H. Townes of U.C. Berkeley, J.W.V. Storey of the Anglo-Australian Observatory, and H.L. Dinerstein, D.F. Lester and M.W. Werner of NASA Ames Research Center. I owe special thanks to Drs. Genzel, Storey and Townes for numerous helpful discussions, to S.C. Beck, S. Beckwith, M.O. Harwit and T.G. Phillips for access to their results in advance of publication, and to B.J. Allen for help in preparing this manuscript. Far-infrared spectroscopy at Berkeley is supported in part by NASA through contract NGR 05-003-511.

References
Far-IR/Submillimeter Receivers

Baluteau, J.P., Anderegg, M., Moorwood, A.F.M., Coron, N., Beckman, J.E., Bussoletti, E. & Hippelein, H.H. 1977, *Appl. Opt.* **16**, 1834. (ESTEC/Meudon Michelson interferometer)

Haller, E.E., Hueschen, M.R. & Richards, P.L. 1979, *Appl. Phys. Lett.*, **34**, 495. (Ge:Ga photoconductors)

Harwit, M., Kurtz, N.T., Russell, R.W. & Smyers, S.D. 1981, *Appl. Opt.* **20**, 3792. (Cornell monochromator-- interferometer receiver)

Houck, J.R. & Ward, D.B. 1979, *Publ. Astron. Soc. Pac.* **91**, 140. (Cornell grating spectrometers)

Phillips, T.G. & Jefferts, K.B. 1973, *Rev. Sci. Inst.*, **44**, 1009. (Bell Labs InSb bolometer heterodyne receiver)

Soifer, B.T. & Pipher, J.L. 1978, *Ann. Rev. Astr. Ap.* **16**, 335. (Review of infrared astronomical instrumentation)

Storey, J.W.V., Watson, D.M. & Townes, C.H. 1980, *Int. J. IR* and *mm Waves*, **1**, 15. (UC Berkeley tandem Fabry-Perot spectrometer)

van Vliet, A.H.F., de Graauw, Th., Lindholm, S. & van de Stadt, H. 1981, *Int. J. IR* and *mm Waves*. **2**, 495. (ESTEC InSb bolometer heterodyne receiver)

Airborne Far-IR/Submillimeter Observations

Baluteau, J.P., Bussoletti, E., Anderegg, M., Moorwood, A.F.M. & Coron, N. 1976, *Ap. J.* (*Letters*) **210**, L45. ([S III], [O III] in M42)

Baluteau, J.P., Moorwood, A.F.M., Biraud, Y., Coron, N., Anderegg, M. & Fitton, B. 1981, *Ap. J.* **244**, 66. ([S III], [O III], [N III] in NGC 7538, W49, M8)

Dain, F.W., Gull, G.E., Melnick, G., Harwit, M. & Ward, D.B. 1978, *Ap. J.* (*Letters*) **221**, L17. ([O III] in various H II regions)

Forrest, W.J., McCarthy, J.F. & Houck, J.R. 1980, *Ap. J.* (*Letters*) **240**, L37. ([O IV], [Ne V] in NGC 7027)

Lester, D.F., Werner, M.W., Storey, J.W.V., Watson, D.M. & Townes, C.H. 1981, *Ap. J.* (*Letters*) **248**, L109. ([O I] in the galactic center)

McCarthy, J.F., Forrest, W.J. & Houck, J.R. 1979, *Ap. J.* **231**, 711. ([S III] in M42, M17, NGC 2024, W51)

Melnick, G., Gull, G.E., Harwit, M. & Ward, D.B. 1978, *Ap. J.* (*Letters*) **222**, L137. ([O III] in M42)

Melnick, G., Gull, G.E. & Harwit, M. 1979, *Ap. J.* (*Letters*) **227**, L29. ([O I] in M42 and M17)

Melnick, G., Gull, G.E. & Harwit, M. 1979, *Ap. J.* (*Letters*) **227**, L35. ([O III] in M42, M17, W51, NGC 6357)

Melnick, G., Russell, R.W., Gull, G.E. & Harwit, M. 1981, *Ap. J.* **243**, 170. ([O I] in NGC 7027)

Moorwood, A.F.M., Balutcau, J.P., Anderegg, M., Coron, N. & Biraud, Y. 1978, *Ap. J.* **224**, 101. ([S III] in M42, W3, NGC 7538)

Moorwood, A.F.M., Salinari, P., Furniss, I., Jennings, R.E. & King, K.J. 1980, *Astr. Ap.* **90**, 304. ([O III], [O I], [N III] in various H II regions)

Moorwood, A.F.M., Baluteau, J.P., Anderegg, M., Coron, N., Biraud, Y. & Fitton, B. 1980, *Ap. J.* **238**, 565. ([S III], [O III], [N III] in M17)

Phillips, T.G., Kwan, J. & Huggins, P.J. 1980, in IAU Symposium No. 87. (Interstellar Molecules), ed. B.H. Andrew (Dordrecht: D. Reidel). (CO, H_2O)

Phillips, T.G., Huggins, P.J., Kuiper, T.B.H. & Miller, R.E. 1980, *Ap. J.* (*Letters*) **238**, L103. ([C I] in various molecular clouds)

Phillips, T.G. & Huggins, P.J. 1981, to appear in *Ap. J.*. ([C I] in various molecular clouds)

Russell, R.W., Melnick, G., Gull, G.E. & Harwit, M. 1980, *Ap. J.* (*Letters*) **240**, L99. ([C II] in NGC 2024 and M42)

Russell, R.W., Melnick, G., Smyers, S.D., Kurtz, N.T., Gosnell, T.R., Harwit, M. & Werner, M.W. 1981, to appear in *Ap. J.*. ([C II] in M17 and NGC 2024)

Storey, J.W.V., Watson, D.M. & Townes, C.H. 1979, *Ap. J.* **233**, 109. ([O III] and [O I] in various H II regions)

Storey, J.W.V., Watson, D.M. & Townes, C.H. 1981a, *Ap. J.* (*Letters*) **244**, L27. (OH in Sgr B2 and Orion-KL)

Storey, J.W.V., Watson, D.M., Townes, C.H., Haller, E.E. & Hansen, W.L. 1981b, *Ap. J.* **247**, 136. (CO in Orion-KL)

Ward, D.B., Dennison, B., Gull, G.E. & Harwit, M. 1975, *Ap. J.* (*Letters*) **202**, L31. (First far-IR line detection: [O III] in M17)

Watson, D.M. & Storey, J.W.V. 1980, *Int. J. IR and mm Waves* **1**, 609. (Review of fine-structure line observations)

Watson, D.M., Storey, J.W.V., Townes, C.H., Haller, E.E. & Hansen, W.L. 1980a, *Ap. J.* (*Letters*) **239**, L129. (CO in Orion-KL)

Watson, D.M., Storey, J.W.V., Townes, C.H. & Haller, E.E. 1980b, *Ap. J.* (*Letters*) **241**, L43. ([O III] in the galactic center)

Watson, D.M., Storey, J.W.V., Townes, C.H. & Haller, E.E. 1981, *Ap. J.* **250**, 605. ([O III], [N III] in various H II regions)

Miscellaneous

Beck, S.C., Bloemhof, E.E., Serabyn, E., Townes, C.H., Tokunaga, A.T., Lacy, J.H. & Smith, H.A. 1982, 253, L83.   *Ap. J. (Letters)*.

Beckwith, S., Evans, N.J. II, Gatley, I., Gull, G.E. & Russell, R.W. 1981, submitted to *Ap. J.*.

Genzel, R. & Downes, D. 1981, Proceedings of the Symposium on Neutral Clouds Near H II Regions: Dynamics and Photochemistry (Dordrecht: D. Reidel).

Goldsmith P.F., Erickson, N.R., Fetterman, H.R., Clifton, B.J., Peck, D.D., Tannenwald, P.E., Koepf, G.A., Buhl, D. & McAvoy, N. 1981, *Ap. J. (Letters)* 243, L79.

Green, S. & Thomas, L.D. 1980, *J. Chem. Phys.* 73, 5391.

Hollenbach, D.J. & McKee, C.F. 1979, *Ap. J. (Supplement)* 41, 555.

Lester, D.F., Bregman, J.D., Witteborn, F.C., Rank, D.M. & Dinerstein, H.L. 1981, *Ap. J.* 248, 524.

McKee, C.F., Storey, J.W.V., Watson, D.M. & Green, S. 1981, submitted to *Ap. J.*.

Oort, J.H. 1977, *Ann. Rev. Astr. Ap.* 15, 295.

Osterbrock, D.E. 1974, *Astrophysics of Gaseous Nebulae* (San Francisco: W.H. Freeman).

Simpson, J.P. 1975, *Astr. Ap.* 39, 43.

THE EXCITATION AND DISTRIBUTION OF CO (J=6→5) EMISSION IN THE ORION NEBULA

D. Buhl[1] and G. Chin[1]
NASA/Goddard Space Flight Center
Greenbelt, MD 20771

G. A. Koepf[1]
Phoenix Corporation
McLean, VA

D. D. Peck[1], and H. R. Fetterman
Lincoln Laboratory, Massachusetts Institute of Technology
Lexington, MA 02173

I. INTRODUCTION

The first detection of the 691 GHz (J = 6→5) transition of CO was made in Orion using the NASA Infrared Telescope Facility (Fetterman et al. 1981) in May 1980. The line detected exhibited a broad, hot emission profile in the BN/KL region of Orion which is also seen in the spectra of several other molecules (HCN, SO_2, SiO, etc.). The narrow line, which dominates the observations of CO at lower frequencies (Phillips et al. 1977 and 1979) was not apparent. The results reported here show a narrow line in both absorption and emission. The intensity of this line shows considerable variation across the small region of the nebula mapped. An analysis of the broad rotationally excited line suggests a very high gas kinetic temperature (T_K>500K) and a hydrogen density of $H_2 \geq 10^6$ mol/cc.

The current observations were made in February 1981 at the Infrared Telescope Facility (IRTF) on Mauna Kea. The spatial CO distribution has been investigated with high resolution (35"). Areas with both narrow line features and broad "plateau" emission have been identified. An attempt has been made to analyze the data quantitatively with a rate equation approach and to fit it to a kinetic model.

---

[1]Visiting Astronomer at the Infrared Telescope Facility which is operated by the University of Hawaii under contract from the National Aeronautics and Space Administration.

II. INSTRUMENTATION AND CALIBRATION

The system used here, (Fetterman et al. 1981; Goldsmith et al. 1981) employs a heterodyne receiver developed for the submillimeter. It consists of a corner cube Schottky diode (Fetterman et al. 1978) serving as the mixer, with a stabilized optically pumped molecular laser (Koepf et al. 1980) serving as the local oscillator. The 1.4 GHz IF amplifier is a FET with a noise temperature of approximately 50 K. The entire system, including an IF matching network had a noise temperature of 3900 K DSB at 432 μm.

The number of 5 MHz channels in the IF filter bank was increased over the earlier experiment to 64 in order to improve the baseline accuracy. Instead of position switching, which was used in the first experiment, the chopping secondary with a 120" arc excursion was used to improve the baseline. The extended distribution of excited CO introduced some reference beam contamination with this approach. This problem will be investigated during a future observing period to determine how severe it is.

During the experiment particular attention was given to calibration of absolute temperatures and to determination of the beam size and shape. Daily measurements were made of the total amount of precipitable water vapor using a Westphal near-infrared solar absorption meter. In addition, atmospheric transmission measurements were made at 691 GHz by recording the atmospheric emission at various elevation angles. This provided the absorption coefficients (τ) at 1 air mass (Fetterman et al. 1981). These measurements were then compared with direct observations of the Moon and Jupiter every few hours. From these comparisons we derived an absorption coefficient for each observing day which varied from τ = 0.6 to 1.2. These values represent extremely low atmospheric absorption at 691GHz and indicate water vapor amounts of <1 precipitable mm. The atmosphere during our observing period was exceptionally good, even for a site like Mauna Kea. All of the observations were corrected using this coefficient, the elevation angle and the measured telescope efficiency.

The absolute intensity scale was derived from repeated observations of Jupiter assuming a continuum temperature of 130°K for the planet (Wright 1976). At the time of observation Jupiter was 40" in

diameter. A map of Jupiter showed that the beam was circularly
symmetric with a relatively low beam efficiency (20%) caused by
spillover of the telescope secondary and led to a determination of 35"
for the beam diameter (FWHM).

III. OBSERVATIONS

Figure 1 illustrates the CO spectra at two points in Orion,
at the KN/BL position and 1' south. These spectra were taken by beam

Fig. 1. CO line profile in the center of the BN/KL region of
Orion and 1' south. The central dip in the line is either
due to self-absorption or a weak emission line in the
reference beam. The profile 1' south shows the classic Orion
shape but has an intensity considerably larger than other CO
lines. This enhanced intensity has also been observed by van
Vliet (1981) in CO (J=4→3).

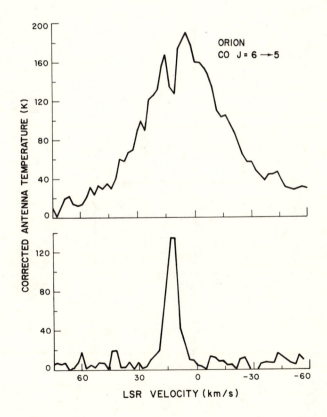

Fig. 2. CO emission (J = 6→5) from Orion as a function of displacement from BN/KL. Along the North-South ridge the results of referencing the chopping secondary in the West direction are also shown. Observations with the reference West are very similar to observations with the reference East suggesting that the profiles have only small amounts of contamination due to CO in the reference beam.

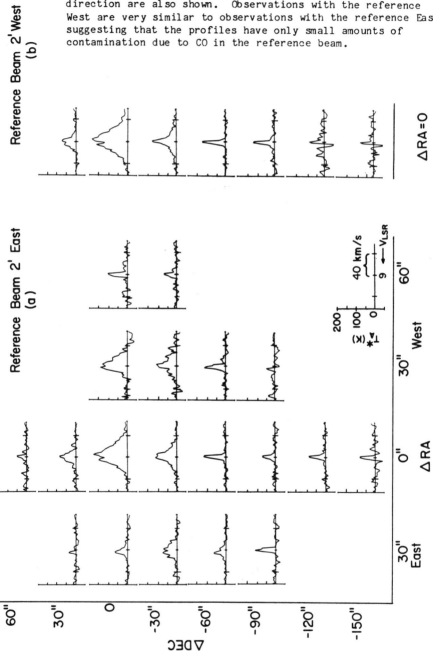

switching the telescope for 6 minutes. Figure 2 shows part of the data
obtained from our observations of Orion as a function of position. The
measurements, which were repeated on different days, indicated that the
intensity scale was consistent to $\pm$ 20%. The mapping of CO (J = 6→5) in
Orion A shows the localized plateau emission (RA (1950) = 5^h 32^m 47^s,
DEC (1950) = -5° 24' 21") in addition to the extended narrow emission
observed at lower frequencies. In comparison with earlier CO
measurements at J=1,2 and 3 there are two apparent differences (see
Figure 2): (i) the narrow emission varies considerably with position
and (ii) the plateau emission is dominant at BN/KL. Most of our antenna
temperatures are not consistent with earlier measurements indicating
that the J=6→5 arises from a hotter region of the cloud. In general
there is an apparent fall-off in intensity away from BN/KL. In addition
to regions of plateau and narrow line emission, there are locations
where both are simultaneously observed such as ΔDEC=-30", ΔRA=0.
Signals appearing with reference West at ΔDEC= -120" and -150" are
interpreted as indicating a differential between CO in the signal and
reference beams. At several positions south of BN we measured
temperatures as high as 120 K. The 3→2 transition of CO has also shown
enhanced emission at BN (Huggins et al. 1981; van Vliet 1981). However,
these hotter regions in the extended spike emission have not been
observed at lower frequencies.

In the center position (BN) we found an antenna temperature
of 180 ± 36 K for the plateau emission, which is somewhat higher than
our previous result (Fetterman et al. 1981; Goldsmith et al. 1981). The
velocity distribution of the earlier measurement of the J = 6→5
transition differs somewhat from our present profile: the FWHM was only
26 ±5 km/s as compared to 38 ±5 km/s, and no dip appeared close to line
center. We conclude from the line profile that the position of the May
1980 spectra was 30" E of the BN/KL peak.

The central absorption spike at BN/KL (Figure 1) may show CO
self absorption due to cooler gas in the outer envelope of Orion.
Another explanation is that the observed dip in the broad emission
feature is subtraction of emission coming from the reference beam. This
implies that the contamination is equal to the absorption T_A^* ∿ 40K. To
try to determine the origin of the absorption feature, observations were
taken along the central north-south ridge which are shown in figure 2

for a reference 2' west of the ridge. Except for ΔDEC-=120" and -150",
the profiles are almost identical, suggesting that the absorption is in
the signal beam and the comtamination is small. Hence, we believe that
CO self absorption is the most likely explanation of the BN/KL spectra.
The lateral extension of the plateau emission, when assumed to have a
Gaussian velocity profile, is about 45" in right ascension and 35" in
declination. This is in close agreement with earlier mappings at lower
J levels (Phillips and Huggins 1977; Solomon, Huguenin and Scoville
1981). The plateau emission, when separated from the narrow feature and
fitted to a Gaussian, is hotter at J = 6→5 than at lower J levels, both
at the center and in the wings. At 25 km/s, almost a perfect J^2 law is
obtained for the antenna temperatures, consistent with an optically thin
line at LTE. At line center the ratio (1→0):(2→1):(3→2):(6→5) is about
1:2:4:20 and can be viewed as the deviation from LTE for regions of
different excitations or evidence of some optical thickness at J=6→5.

IV. DISCUSSION

The kinetic processes that cause the hot plateau emission in
Orion have been the subject of extensive speculation and modeling. The
most widely applied models are based on large scale velocity gradient
structures (Goldreich and Kwan 1974; Scoville and Solomon 1974). The J
= 3→2 and 2→1 plateau profiles have been matched with a differential
expansion model by Kwan and Scoville (1976) and by Wannier and Phillips
(1977), respectively. Our J = 6→5 profile can be fitted to this model
only by assuming a substantially higher velocity gradient. The
detection of molecular hydrogen emission has recently led to the model
of a shock wave expanding into the molecular medium (Kwan 1977). For
conservation of the H_2 molecules in the hot shock front, a maximum shock
velocity of 24 km/s has to be imposed. The detection of CO emission
from high J levels (Watson et al. 1980) has also been attributed to such
post-shock regions where the kinetic temperature has decreased to about
1000 K (Storey et al. 1981). Unfortunately, these observations lack the
spectral resolution required to give insight into the velocity details
and the possible shell structure of the emission regions.

The J = 6→5 emission could have its origin in an even cooler
region farther behind such a shock front. In this region, however, the
gas velocity has also dropped to substantially below its value at the

shock front, and it is difficult to explain the width of the plateau
emission. Hollenbach and McKee (1979) have pointed out that
considerably faster shocks that produce molecular dissociation are a
possibility. At some distance behind the front of such fast shocks the
temperature has decreased because of efficient cooling mechanisms, to
values that allow the re-formation of the pre-shock molecular
constituents while the gas is still moving at a high velocity. An
alternative explanation of the high velocity emission given by Chevalier
(1980) assumes the ejection of knots of gas at high velocities up to 100
km/s for an active source, possibly the BN object. As these knots move
out into the molecular cloud, shock waves with a large range of
velocities are generated that can account for the observed H_2 and CO
emission spectra. Modeling of the observed high resolution lineshape
requires a number of assumptions for the density, temperature and
velocity structure as well as the shock geometry.

Although this problem will be treated in detail in a future
paper, we can approach the analysis by assuming the line to be optically
thin and therefore a large excitation temperature, $T_{ex} >> 180^{\circ}$K. This is
supported by the scaling of the measured plateau CO transitions in the
wings approximately as J^2. Taking our value of T = 180 K and FWHM of 38
km/s for the CO (6→5) transition yields a total CO column density of 3.2
x 10^{17}/cm^2. The antenna temperature can now be calculated for a limited
set of J values by solving the statistical equilibrium equations which
relate the fractional population of the various rotational levels to the
rate in which collisions and radiative transitions populate these
rotational levels. The collision rate constants for CO-He were obtained
from a program provided by Green (Green and Chapman 1978). We scaled
the CO-He values by 1.5 to obtain the CO-H_2 collision constants from
Green's program. For various orders of H_2 density and for kinetic
temperatures of less than 500 K, the first 25 levels were sufficient to
describe the fractional population of CO (Watson et al. 1980). The
calculation is constrained such that the J = 6 level has the column
density required to yield our measured value of 180 K. The column
density of the other J levels were scaled by the fractional population
given by the solution to the statistical equilibrium for the various
kinetic temperatures at a H_2 density of $\sim 10^6$cm^{-2}.

Examining the results of this calculation (Figure 3), a

kinetic temperature of 100 K was found to be an unrealistic value for
our set of parameters. However, for kinetic temperatures of 200 K and
higher the cloud model yields temperatures which correspond within
experimental errors to the measured antenna temperatures. The most
severe departure of the calculated values from measured data occurs for
the J = 3→2 transition, with a predicted antenna temperature in excess
of 48 K, compared with a measured value of 35 K. Measurement of the
next higher rotational level, CO J = 7→6, using our instrument with

Fig. 3. A model of CO emission temperatures as a function of
J for various kinetic temperatures at a hydrogen density of
$H_2=10^6$ mol/cm^{-3}. The curves were forced to coincide with our
measurement of the J=6→5 transition. A hot, optically thin
cloud is required to fit the data points (+). The model is
isothermal, without any external infrared radiation. It also
assumes that all transitions originate from the same
environment. This last assumption may not be justified.

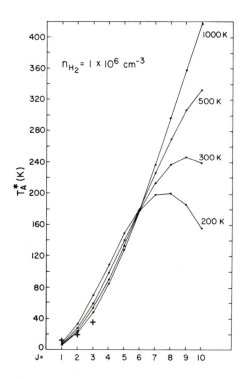

different laser L.O. lines, will provide a better estimate of the actual
kinetic temperature of the excited gas.

Assuming a H_2 density of 10^6 and a mixing ratio of $X(CO)$ =
10^{-4} - 10^{-5}, the path length over which the CO 6→5 transition occurs is
less than a parsec. This indicates that the emitting region for the
plateau emission is a wide thin sheet very similar to the region in
which the H_2 emission arises. The similarity between the H_2 emission
and the central profile obtained at BN-KL also extends to the
asymmetrical profile which is apparent in our 6→5 observation and also
to the resolved H_2 profiles obtained by Nadeau and Geballe (1979). It
appears reasonable, therefore, that the CO is also excited in a region
after a shock (Watson et al. 1980; Storey et al. 1981). A complete
modeling of the CO profiles at various J levels can therefore be
attempted by assuming that the various velocity components which arise
after a shock have different local density and temperature values. A
statistical equilibrium model for the optically thin medium can then be
solved for each point in the velocity profile. In solving for the
fractional population of CO at high temperatures and densities it must
be noted that many of the rotational levels ($J \leq 6$) appear to be
strongly inverted at large ranges of temperatures and densities (McKee
et al. 1982). Since the optical depth is small, however, these effects
are not expected to be significant in this particular set of
measurements.

V. CONCLUSION

These observations have demonstrated that high resolution
mapping is possible at submillimeter wavelengths. The distribution of
CO radiating in the $J = 6→5$ transition in Orion has been shown to be
composed of two separate regions. There is a core approximately 45" in
size with a hot broad plateau emission. Outside of this region the
emission is due to strong narrow line sources.

New models are required to explain the detailed observations
which must be fitted for both the hot broad (180K) and narrow line
(120K) regions. In addition, measurements of higher J levels are needed
to narrow the range of solutions possible for the $J = 6→5$, 3→2, 2→1 and
1→0 data.

The sensitivity of the laser heterodyne receiver with good

atmospheric conditions is now well suited to making observations of molecular clouds. Our preliminary data indicate that these high resolution measurements can give valuable insight into distribution of molecules and to the dynamic processes present in pre-main sequence stellar regions. Used in conjunction with lower frequency measurements, these submillimeter results can now make important contributions to our quantitative understanding of molecular interactions in stellar environments.

VI. ACKNOWLEDGEMENT

We wish to thank C. J. Peruso, C. L. Lowe, T. Kostiuk and C. D. Parker for support of this experiment. We also acknowledge the leadership of N. McAvoy in the submm laser program during the past several years. Support for various parts of this experiment from B. J. Clifton and P. E. Tannenwald are appreciated, as well as the helpful comments of S. G. Kleinmann in preparing this paper. Finally we wish to express our appreciation to E. E. Becklin and the staff of the IRTF for their enthusiastic support of this experiment.

References

Chevalier, R. A. (1980), On the High Velocity Gas Motions in the Orion
        Molecular Cloud. Astrophys. Lett., 21, 57-61.
Erickson, N.R. (1979). Ph.D. Thesis, Physics Dept., University of
        California, Berkeley.
Fetterman, H. R., Tannenwald, P. E., Clifton, B. J., Parker, C. D.,
        Fitzgerald, W. D. Erickson, N. R. (1978). Far-IR Heterodyne
        Radiometric Measurements with Quasi optical Schottky Diode
        Mixers. App. Phys. Let., 33, 151-154.
Fetterman, H. R., Koepf, G. A., Goldsmith, P. F., Clifton, B. J.,
        Buhl, D., Erickson, N. R., Peck, D. D., McAvoy, N., &
        Tannenwald, P. E. (1981). Submillimeter Heterodyne Detection
        of Interstellar Carbon Monoxide at 434 Micrometers. Science,
        211, 580-582.
Goldsmith, P. F., Erickson, N. R., Fetterman, H. R., Clifton, B. J.,
        Peck, D. D., Tannenwald, P. E., Koepf, G. A., Buhl, D. &
        McAvoy, N., (1981). Detection of the J=6→5 Transition of
        Carbon Monoxide. Ap. J. (Letters), 243, L79-L82.
Goldreich, P., & Kwan, J. (1974). Molecular Clouds. Ap. J., 189,
        441-453.
Green, S., & Chapman, S. (1978). Collisional Excitation of
        Interstellar Molecules: Linear Molecules CO, CS, OCS, and
        HC_3N. Ap. J. Suppl., 37, 167.
Huggins, P. J., Phillips, T. G., Blair, G. N. & Solomon, P. M. (1981).
        Detection of ^{13}CO (J=3-2) Emission from the Molecular Clouds
        OMC-1 and NGC 2264. Ap.J., 244, 863-868.
Hollenbach, D., & McKee, C. F. (1979). Molecule Formation and
        Infrared Emission in Fast Interstellar Shocks: I. Physical
        Processes. Ap. J., Suppl., 41, 555.
Koepf, G. A., Fetterman, H. R. & McAvoy, N. (1980). A Stable Sub-mm
        Oscillator for Heterodyne Radiometry and Spectroscopy. Int.
        J. for Infrared and Millimeter Waves, 1, 597-607.
Kwan, J., & Scoville, N. Z. (1976). The Nature of the Broad Molecular
        Line Emission at the Kleinmann Low Nebular. Ap. J.
        (Letters), 210, L39-L43.
Kwan, J. (1977). On the Molecular Hydrogen Emission at the Orion
        Nebula. AP. J., 216, 713-723.
McKee, C. F. Storey, J. W. V., Watson, D. M. & Green, S. (1982).
        Far-Infrared Rotational Emission by Carbon Monoxide. Ap.J.
        (submitted).
Nadeau, D. & Geballe, T. R. (1979). Velocity Profiles of the 2.1
        Micron H_2 Emission Line in the Orion Molecular Cloud. Ap. J.
        (Letters), 230, L169-L173.
Phillips, T. G. & Huggins, P. J. (1977). Observations of Carbon
        Monoxide J=2-1 Isotopic Lines in DR21, W51 and Orion. Ap.
        J., 211, 798-802.
Phillips, T. G., Huggins, P. J. Neugebauer, G. & Werner, M. W. (1977).
        Detection of Submillimeter (870 μm) CO Emission from the
        Orion Molecular Cloud. Ap. J. (Letters), 217, L161-L164.
Phillips, T. G., Huggins, P. J., Wannier, P. G. & Scoville, N. Z.
        (1979). Observations of CO (J=2-1) Emission from Molecular
        Clouds. Ap. J., 231, 720-731.
Plambeck, R. L. & Williams, D. R. W. (1979). Hydrogen Densities in

Molecular Clouds Inferred from CO Observations. Ap. J. (Letters), 227, L43-L47.

Scoville, N. Z. & Solmon, P. M. (1974). Radiative Transfer Excitation and Cooling of Molecular Emission Lines (CO and CS). Ap. J., (Letters), 187, L67-L71.

Solomon, P. M., Huguenin, G. R. & Scoville, N. Z., (1981). The Source of High Velocity Emission at the Orion Molecular Cloud Core. Ap. J. (Letters), 245, L19-L22.

Storey, J. W. V., Watson, D. M., Townes, C. H., Haller, E. E., & Hansen, W. L. (1981). Far-Infrared Observations of Shocked CO in Orion. Ap. J., 247, 136-143.

Van Vliet, A. H. F. (1981). A Submillimeter Heterodyne Receiver and its Application in Astronomy. Ph.D. Thesis, University of Utrecht, Netherlands.

Wannier, P. G. & Phillips, T. G. (1977). Evidence for Optically Thin CO Emission from the Kleinmann-Low Nebula. Ap. J., 215, 796-799.

Watson, D. M., Storey, J. W. V., Townes, C. H., Haller, E. E., Hauser, W. L. (1980). Detection of CO J=21→20 (124.2 μm) and J=22→21 (118.6 μm) Emission from the Orion Nebula. Ap. J. (Letters), 239, L129-L132.

Wright, E. L. (1976). Recalibration of the Far-Infrared Brightness Temperatures of the Planets. Ap. J., 210, 250-253.

Zuckerman, B., Kuiper, T. B. H. & Kuiper E. N. R. (1976). High Velocity Gas in the Orion Infrared Nebula. Ap. J. (Letters), 209, L137-L142.

FIRST OBSERVATIONS OF THE J = 4 → 3 TRANSITION OF HCO$^+$.

Rachael Padman
Mullard Radio Astronomy Observatory, Cavendish Laboratory, Madingley Road,
Cambridge, England.

Paul Scott
Mullard Radio Astronomy Observatory, Cavendish Laboratory, Madingley Road,
Cambridge, England.

Adrian Webster
Mullard Radio Astronomy Observatory, Cavendish Laboratory, Madingley Road,
Cambridge, England.

Abstract. The J = 4 → 3 transition of HCO$^+$ was observed in
OMC1 and in three other galactic sources which have self-
reversed CO profiles. None of the sources shows self-
absorption in this line and, in NGC2071 and Mon R2, the line
wings seen in CO and in HCO$^+$ 1 → 0 have disappeared. Minimum
masses between 200 and 1200 $M_\odot$ are derived for the high-density
cores, and evidence is found for a new, compact, optically-thin
source in OMC1. The rest frequency for the HCO$^+$ 4 → 3 transition is found to be 356.7344 ± 0.0005 GHz, giving values of
44,594.44 ± 0.014 MHz and 82.0 ± 2.4 kHz for the molecular
constants B_0 and D_0.

The observations were made on 1981 January 16 with the InSb
hot-electron mixer receiver of the 3.8 m UKIRT on Mauna Kea, Hawaii; on-
and off-source measurements were interleaved in the usual way, were calibrated against absorbers at known temperatures and were corrected for
atmospheric attenuation by sky-dipping. The local oscillator was set at
a frequency derived from the molecular constants of Huggins et al. (1979),
but the transition was found consistently to lie $\sim$ 1 km s^{-1} to lower
velocity; we have therefore revised the transition frequency and molecular
constants to the values given in the abstract.

In Fig. 1 is shown a spectrum of OMC1, smoothed to an effective
resolution of 1.7 km s^{-1}. There is a strong feature at $v_{LSR} \sim$ 9 km s^{-1},
a velocity typical of many molecules in this source, but it has an
unexpected extension or blend on the high-velocity wing. The whole
feature can conveniently be approximated by two Gaussians, as shown, with
the new component centred on 13 km s^{-1}. This component is not prominent
in spectra of the lower transitions of HCO$^+$, so we suppose it to be
optically thin and hot; there is some corresponding broadening of the
3 → 2 spectrum of Huggins et al. (1979) compared with that of Erickson
et al. (1980) made with a broader beam, so the component may also be

Fig. 1.   HCO⁺ J = 4 → 3 spectra taken with a beam of 1 arcmin.

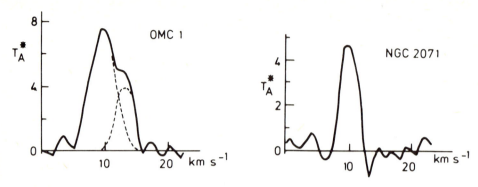

compact.   The nature of the new emission feature is not clear, but it could be related to the 12.6 km s⁻¹ feature observed in CO 1 → 0 to the NW of BN by Loren (1979), which would appear to be a continuation of a partial ring of high-velocity gas identified as the shock associated with an ionisation front.   The shock could provide the excitation for the new HCO⁺ component and could also substantially increase the relative abundance of this molecule (Dickinson et al. 1980).

Observations were also made of GL961, NGC2071 and Mon R2. These sources were selected because they all show interesting self-absorption features, either in HCO⁺ 1 → 0 or in CO;   the emission profiles of the two latter sources also have high-velocity wings.   Fig. 1 shows our result for NGC2071, which is typical of the others.   The self-absorption and the emission wings are nowhere to be seen, and all that remains is a strong and relatively narrow emission feature, presumably from the dense core of this object.   A likely reason for the disappearance of the wings is that the excitation is insufficient due to the density being too low; a density of H_2 in the range $3 \times 10^4 - 1 \times 10^6$ cm⁻³ would be suitable to excite the 1 → 0 transition but not the 4 → 3.   As for the core, a minimum mass may be obtained from the smallest number of HCO⁺ molecules in the J = 4 state necessary to give the observed emission and an assumed value for the HCO^+/H_2 abundance ratio (3×10^{-11}, Loren & Wootten 1980).   Minimum masses in the range 200-1200 $M_\odot$ are found, with the principal uncertainty stemming from the assumed abundance ratio: it is not clear to us that this standard value for the more tenuous parts of the clouds is appropriate to the dense cores.

References

Dickinson, D.F., Rodriguez-Kuiper, E.N., Onger, A.St.Clare & Kuiper, T.B.H.,
        1980.   Astrophys. J., 237, L43.
Erickson, N., Davis, J.H., Evans, N.J., Loren, R.B., Mundy, L., Peters, W.L.,
        Scholtes, M. & Vanden Bout, P.A., 1980.   IAU Symp. No.87, 25.
Huggins, P.J., Phillips, T.G., Neugebauer, G., Werner, M.W., Wannier, P.G.
        & Ennis, D., 1979.   Astrophys. J., 227, 441.
Loren, R.B., 1979.   Astrophys. J., 234, 2207.
Loren, R.B. & Wootten, A., 1980.   Astrophys. J., 242, 568.

OBSERVATIONS OF THE J=2→1 CO LINE IN MOLECULAR CLOUDS SHOWING
AMMONIA EMISSION

L.T. Little
Electronics Laboratory, University of Kent at Canterbury.

Abstract. Observations of the J=2→1 transitions of ^{12}CO and
^{13}CO at 230 and 220 GHz in 18 molecular clouds have been made
at the UK Infrared Telescope. The sources were chosen on the
basis of their ammonia emission. A comparison between ^{12}CO
and ^{13}CO spectra reveals a variety of absorption effects,
ranging from slight asymmetries in the ^{12}CO profiles relative
to their ^{13}CO counterparts (W43S, S88) to a deep self
absorption dip (S68 and possibly G34.3+0.1). The asymmetry
observed in nine sources out of eighteen is most easily
explained if the clouds are collapsing; there is no clear
evidence for expansion. The ^{13}CO linewidths are systematically wider than those from the NH_3 cores, suggesting that
the velocity dispersion in the sources increases with
distance from the centre.

1 INTRODUCTION

Observations of higher-order transitions, $J(>1) \to J-1$, of
carbon monoxide are of importance since the large optical depths and
smaller beamwidths obtainable at these frequencies facilitate observation
of self-absorption effects (e.g. Snell & Loren 1977; Kwan 1978; Leung
1978) which may prove indispensable in the determination of the velocity,
density and temperature gradients within molecular clouds. One approach
towards understanding the velocity fields within molecular clouds is to
compare the line-width and angular distribution of a high excitation
molecular transition with transitions of CO which needs significantly
lower densities. From such observations, some idea of how the velocities
within the clouds may vary with distance from the centre is obtained, but
their sense (expansion or collapse) and the ratio of turbulent to
systematic velocity remain uncertain. Observations of higher-order
transitions of CO are pertinent to the latter two questions. The very
high optical depth of ^{12}CO may produce strong absorption and asymmetrical
line shapes whose sense of skew is an indicator of collapse or expansion.
To distinguish whether such asymmetries are the results of these effects,

or due to genuine differences in matter distribution with velocity, observations of the less abundant isotope ^{13}CO are vital. The problem of self-absorption in CO spectra has been addressed by several authors. Snell & Loren (1977) considered that the relation of the prominent self-absorption dip in Mon R2 to its ^{13}CO emission peak represented evidence for collapse of the source with a velocity which increased towards the centre. On the other hand, Kwan (1978) was able to interpret the relatively small asymmetry of the ^{12}CO spectra of M17 compared to its ^{13}CO counterpart by assuming the source to consist of a turbulent core with a collapsing outer envelope whose velocity increases from the centre (V ∝ r). Although several sources showing absorption dips have now been discovered, collectively they present no clear evidence for either collapse or expansion, or indeed whether the absorbing sources are closely tied to the emission regions (see e.g. Phillips et al. 1981, Loren et al. 1981).

To widen the search for self-absorption effects a suitable sample of dense sources is required. Over the last three years a search at the Chilbolton Observatory for ammonia emission from over 200 candidate molecular clouds with δ > -10° (Macdonald et al. 1981) has yielded 36 new detections. Since ammonia molecules are relatively rare and require densities $\geq 10^4$ cm^{-3} for detection, these new ammonia sources represent a most promising collection of dense regions for study. Eight of them have been mapped in NH₃ and most appear to contain clumpy cores (e.g. Little et al. 1979, 1980; Brown et al. 1981). In this paper J=2→1 CO observations are presented for a number of these sources and are compared with ammonia and hydroxyl results. The CO observations were obtained during two observing sessions at the 3.8 m UK Infrared Telescope, Mauna Kea, using a 230 GHz uncooled Schottky barrier diode mixer and a 512 channel digital auto-correlation spectrometer. The receiver, and many of the observations, are described by Riley et al. 1982. The sources found in the Kent ammonia survey fall into two classes, namely those associated with (i) optical nebulosities (mostly Herbig-Haro objects from the list given by Gyulbudaghian, Glushkov & Denisyuk 1978), and (ii) prominent radio H[+] emission. The former are generally closer (distances ~1 kpc) than the latter (distances ~ a few kpc). Owing to their disposition on the sky, all but two of the sources observed were of the second class; the exceptions were S68, an optical nebulosity which contains several very compact H[+] regions, and R131.

Figure 1(a) ^{12}CO(—) and ^{13}CO(- - -) spectra showing relative asymmetry. The LSR velocities are marked relative to the NH₃ velocity. The ^{12}CO tends to be displaced to lower velocities compared to NH₃ and ^{13}CO.

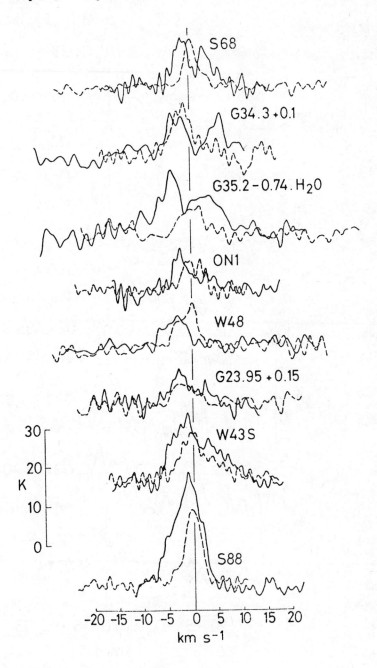

Figure 1(b) ^{12}CO(—) and ^{13}CO(---) spectra with little measurable relative asymmetry. LSR velocities are marked relative to the NH₃ velocity.

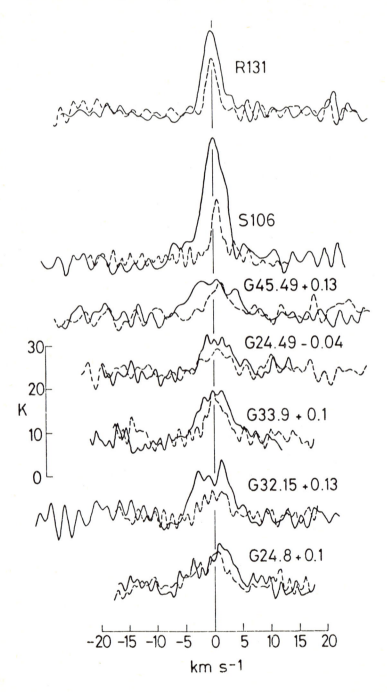

2   OBSERVATIONS

Selected spectra of the J=2→1 ^{12}CO and ^{13}CO transitions, smoothed to 0.6 kms^{-1} resolution, are shown in figure 1.   The velocities are believed to be correct to within 0.2 kms^{-1}.   At 230 GHz the telescope has a beamwidth 1.25 arcmin and a beam efficiency ~0.5.   Antenna temperatures have been corrected for atmospheric losses and divided by the beam efficiency to obtain the corrected Rayleigh-Jeans beam brightness temperature.

3   ANALYSIS

3.1   Radial variation of velocity

On the basis of a simple two-level model for line excitation, it is easy to show that for similar optical depths the densities required to excite metastable inversion transitions of NH$_3$ are approximately 15 times greater than those for J=1→0 ^{13}CO.   Thus it is no surprise to find that the angular extent of J=1→0 ^{13}CO emission is often greater than that of NH$_3$ (Little et al. 1979, 1980).   The effects of enhanced radiative trapping due to the greater optical depth of the J=2→1 ^{13}CO transition and line interlocking (Leung 1978) are likely to lead to a near-equalization of the excitation densities for the J=2→1 and J=1→0 ^{13}CO transitions, and it is thus reasonable to expect the angular distribution of J=2→1 ^{13}CO to be similar to that of J=1→0.   Recent observations of S68 and G35.2-0.74 (Little et al., in preparation) indicate that the J=2→1 ^{13}CO emission is indeed more extended than that of ammonia.

Figure 2(a) shows a plot of the NH$_3$ line-widths versus those of the J=2→1 transitions of ^{13}CO.   It is evident that there is a strong correlation, and that the NH$_3$ line-widths are systematically less ($\sim\frac{1}{2}$) than those of ^{13}CO.   The interpretation of these facts is complicated by:

1. Line saturation.   The difference of line-widths could be explained if the ^{13}CO optical depth were much greater than that of NH$_3$. The optical depths of the NH$_3$ lines are known from their hyperfine ratios to be not much greater than unity: $\tau(^{13}CO)$ is unknown.   The spectral broadening by saturation will be negligible either if $\tau \ll 1$, or if $T_{EX} \ll T_K$ and systematic velocity gradients dominate over turbulence. The relative brightness of the ^{13}CO lines compared to ^{12}CO strongly suggests that neither of these possibilities is true and therefore that saturation broadening is significant.   A value for $\tau(NH_3) \sim 1$ implies that $\tau(^{13}CO) \gtrsim$

30 if the velocity dispersion of CO, corrected for saturation broadening, is to be less than that of NH_3; this will be the case for velocity dispersions that increase towards the centre of the sources.

2. Different antenna beamwidths. Velocity gradients (e.g. rotation or shear) across the line of sight would tend to produce larger widths in the lines observed with the largest telescope beams, i.e. a larger line-width for NH_3 (beamwidth 2.2 arcmin) than for CO (beamwidth 1.1 arcmin). The observed effect is the opposite of this.

Since the ^{13}CO observations were made with a narrower beamwidth than those of NH_3, and yielded larger linewidths, it would be possible to argue that the linewidths increase towards the centre of the sources. However it was suggested above that in fact the ^{13}CO arises from a more diffuse region than the NH_3 cores, in which case the reverse is correct. At this point it is instructive to make a comparison with observations of the 18 cm OH lines which, like ^{13}CO, probably requires lower excitation densities than NH_3, but have been observed with a wider beam.

Observations of 16 of the Kent NH_3 sources have already been made in weak anomalous 18 cm lines of OH with the Nancay radiotelescope (Little & Cesarsky, 1981). Although both CO and OH are excited at relatively low densities, the OH observations are made with a large beamwidth ($\sim 4 \times 24$ arcmin^2) compared to those of ^{13}CO, and there is no obvious reason why the saturation of the OH lines should be similar to that of ^{13}CO. Also the 1612 MHz emission lines in the OH survey are probably weakly inverted but have similar widths to the 1720 MHz absorption lines, whereas saturation effects might have been expected to modify the linewidths in opposite senses. A comparison of the $\Delta V(NH_3)$ vs $\Delta V(OH)$ plot with that of $\Delta V(NH_3)$ vs $\Delta V(^{13}CO)$ is therefore of interest, and the former is presented in figure 2b. The OH plot contains sources associated both with optical nebulosities and with radio compact H$^+$ regions. Otherwise the OH and CO plots are of rather similar form, and the NH_3 line-widths are $\sim 2/3$ those of OH. Although it is possible that some unfortunate combination of saturation, beamwidth and excitation-density effects for OH and CO has produced the similarity, it is more likely that the results may be taken as indicating that (i) velocity gradients across the line of sight are not a predominant factor in determining the line-widths (since similar widths are obtained for lines of OH and CO which are observed with

Figure 2 (a) NH$_3$(1,1) linewidths plotted against ^{13}CO J=2→1
linewidths.   (b) NH$_3$(1,1) linewidths plotted against OH line-
widths.   • strong radio H$^+$ region; × optical nebulosity;
o cold dark cloud.

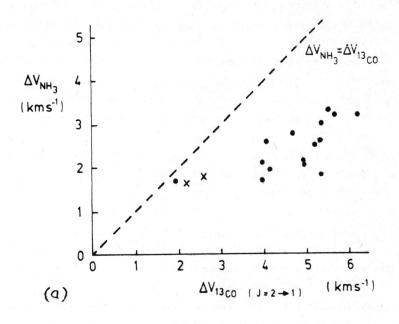

(a)

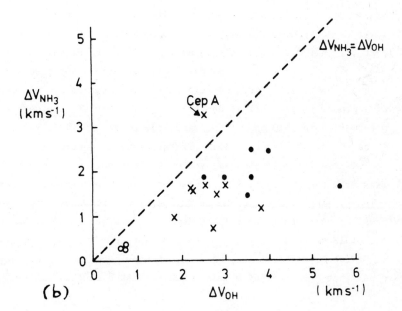

(b)

very different beamwidths), and hence (ii) velocities (whether systematic
or turbulent) in the NH_3 cores are genuinely less than those in the outer
envelopes (as evidenced by CO and OH).

If true, this has a further significant consequence. As noted
by Little & Cesarsky, there appears to be a systematic increase in the
observed line-widths in progressing from the cold dark clouds through
those displaying Herbig-Haro objects, to the clouds associated with
prominent compact H^+ regions. Since this is also an order of increasing
distance, the effect could have been explained by the presence of velocity
gradients across the line of sight, as for a constant beamwidth these
would have a more marked effect on the line-widths of more distant sources
(provided that they are resolved). The argument of the preceding paragraph suggests that velocity gradients are not in fact significant so that
the trend of increasing line-width from cold cloud to H^+ region is
intrinsic to the molecular clouds themselves.

3.2 Expansion or contraction

A comparison between the ^{12}CO and ^{13}CO spectra in figure 1
reveals a wide range of self absorption effects, ranging from slight
asymmetries in the ^{12}CO spectra relative to their ^{13}CO counterparts
(enhancements to the low velocity side in W43S and S88) to deep self
absorption dips (S68 and probably G34.3+0.1). ON1 and W48 both show a
broad self-absorption with a marked displacement of the ^{12}CO maximum to
lower velocity compared with the ^{13}CO or NH_3. It is possible, though not
certain, that G24.8+0.1 is similarly displaced to higher velocities.

It is well known that radiation from a hot core will be
absorbed by material at lower excitation temperatures in the outer regions
of a cloud, if its optical depth is significant. This effect will be much
more marked in ^{12}CO than ^{13}CO, since the former is ≥ 40 times as abundant,
as can be clearly seen in S68. If the cloud is contracting, the outer
regions along the line of sight from the hot core will tend to be moving
away from the observer, so the absorption will be greater in the high-
velocity wing. If the cloud is expanding the sense will be reversed.
Thus the shift to lower velocity of ^{12}CO relative to ^{13}CO observed in 9
out of 18 sources is evidence for contraction. The opposite displacement
of G24.8+0.1, if verified, would provide evidence for expansion. It is
clear that a variety of models could be fitted to the limited data. A

model by Kwan (1978) produced slight asymmetries not dissimilar to those observed in S88 and W43S. This model involved a gradual transition from a turbulent region at the centre of the cloud to a systematic motion at the edge, a slow decline in temperature and a rapid decline in density with radius ($\propto r^{-3}$) from a peak value $H_2 \sim 2{\times}10^5$ cm^{-3}. A greater optical depth in the outer absorbing material could well explain the spectra of W48 and ON1, while a slightly reduced velocity dispersion could probably give rise to the absorption dips seen in S68, and possibly G34.3+0.1 and G35.2-0.74.

4   CONCLUSIONS

Taken in conjunction with complementary NH_3 and OH results, our CO observations of the molecular clouds associated with compact H^+ regions provide evidence in favour of the following assertions:

> (1) the velocity dispersion in the ammonia cores is less than that in the surrounding envelopes;

> (2) velocity gradients across the line of sight are not generally significant in determining the line widths;

> (3) there is a systematic increase in velocity dispersion in going from cold dark clouds (in which there is no recent evidence of star formation), through Herbig-Haro objects (near recently formed stars) to the well-developed radio compact HII regions;

> (4) there is evidence for contraction in an appreciable number of the sources, 9 out of 18, but none certain for expansion.

Stenholm et al. (1981) have applied detailed radiation transfer calculations to fit models to the average profiles of CO, CS and HCO^+ emission derived for a sample of molecular clouds probably at a rather earlier stage of their evolution than the radio-powerful H^+ regions which predominate in our observations. They find a good fit for a model cloud which has a density varying nearly as r^{-2}, with a systematic velocity of collapse and a turbulent velocity both of similar magnitude and increasing slowly towards the cloud centre. This would suggest a velocity dispersion which increased towards the cloud centre, as does not appear to be the case for our sample of sources, given the validity of the rather simple arguments presented above. It would be of interest to apply the

theory of Stenholm <u>et al</u>. to our observations to determine what modification to our conclusions it would require.

Acknowledgements

Many of the observations and arguments in this paper have been
presented in Riley <u>et al</u>. (1982). Other CO observations were made on an
observing trip to UKIRT by A.T. Brown, G.H. Macdonald, P.W. Riley,
D. Vizard, and the author.

References

Brown, A.T., Little, L.T., Macdonald, G.H., Riley, P.W. & Matheson, D.N.
(1981). Mon. Not. R. astr. Soc., <u>195</u>, 607.

Gyulbudaghian, A.L., Glushkov, Yu. I. & Denisyuk, E.K. (1978).
Astrophys. J., <u>224</u>, L137.

Kwan, J. (1978). Astrophys. J., <u>223</u>, 147.

Leung, C.M. (1978). Astrophys. J., <u>225</u>, 427.

Little, L.T. & Cesarsky, D.A. (1981). Astron. Astrophys., submitted.

Little, L.T., Macdonald, G.H., Riley, P.W. & Matheson, D.N. (1979). Mon.
Not. R. astr. Soc., <u>188</u>, 429.

Little, L.T., Brown, A.T., Macdonald, G.H., Riley, P.W. & Matheson, D.N.
(1980). Mon. Not. R. astr. Soc., <u>193</u>, 115.

Loren, R.B., Plambeck, R.L., Davis, J.H. & Snell, R.L. (1981).
Astrophys. J., <u>245</u>, 495.

Macdonald, G.H., Little, L.T., Brown, A.T., Riley, P.W., Matheson, D.N.
& Felli, M. (1981). Mon. Not. R. astr. Soc., <u>195</u>, 387.

Morris, M., Zuckerman, B., Palmer, P. & Turner, B.E. (1973). Astrophys.
J., <u>186</u>, 501.

Phillips, T.G., Knapp, G.R., Huggins, P.J., Werner, M.W., Wannier, P.G.,
Neugebauer, G. & Ennis, D. (1981). Astrophys. J., <u>245</u>, 512.

Riley, P.W., Little, L.T., Brown, A.T., Hills, R.E., Padman, R.,
Vizard, D., Lesurf, J.C.G. & Cronin, N.J. (1982). Mon. Not.
R. astr. Soc., <u>199</u>, 197.

Snell, R.L. & Loren, R.B. (1977). Astrophys. J., <u>211</u>, 122.

Stenholm, L.G., Hartquist, T.W. & Morfill, G.E., (1981). Astrophys. J.,
<u>249</u>, 152.

OBSERVATIONS OF CIRCUMSTELLAR SHELLS AROUND OH/IR STARS

R.S. Booth, R.P. Norris & P.J. Diamond
Nuffield Radio Astronomy Laboratories,
Jodrell Bank, Macclesfield, Cheshire

Abstract   Interferometric observations of the 1612 MHz OH maser emission from late type stars has revealed circumstellar shells of gas some 10^3-10^4 A.U. in diameter. Using the velocity information of the OH spectrum and an estimate of the molecular density it has been possible to derive mass loss rates from these stars which are of the order of $10^{-5}M_\odot$ - $10^{-4}M_\odot$ per year.

Introduction

It is well known that some late type stars exhibit maser radiation from the molecular species SiO (Silicon monoxide), H_2O (Water) and from the hydroxyl radical, OH. These stars are cool oxygen rich giants or supergiants usually of spectral type M5 or later, they are also strong IR sources. The properties of the masers have been reviewed by Elitzur (1981).

The OH emission normally associated with the late type stars is from the ground state ($^2\Pi_{3/2}$, $J = 3/2$) Λ-doublet at frequencies of 1612 MHz (F=1→2), 1665 MHz (F=1→1) and 1667 MHz (F=2→2). The observations reported in this paper are of the strong 1612 MHz emission from two OH/IR stars OH127.8-0.0 and OH104.9+2.4. Both were discovered in a survey by Bowers (1978) and have since been identified with IR objects through accurate position measurements (Bowers et al. 1980; Reid et al. 1977; Porter et al., in preparation).

The 1612 MHz spectra of OH127.8-0.0 and OH104.9+2.4 are reproduced in Figs. 2 and 4. Both show two main regions of emission separated by 20-30 km s^{-1}, a signature which is characteristic of objects of this type and which has led to the suggestion that the OH lies in an expanding shell surrounding the star.

The Observations

These observations of the spatial distribution of the OH maser emission were carried out with three or more telescopes of the new Jodrell Bank Multi-Element Radio Linked Interferometer Network (MERLIN). The network consists of six telescopes located at sites surrounding Jodrell Bank separated by distances up to 134 km and which are indicated on the diagram of Fig. 1.

Each outstation telescope is remotely controlled from Jodrell Bank and microwave links carry the radio signals back to Jodrell Bank. Local oscillators at all sites are phase locked through a communications

Fig.1   Diagram indicating the locations of the telescopes and the baselines of the Jodrell Bank Multi element radio linked interferometer network (MERLIN)

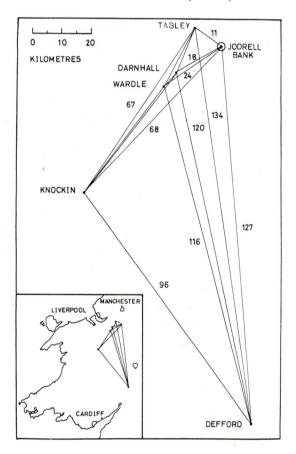

link. The link path delays are continuously measured and corrections for
their variation made to the phase of the received signals.

　　　　　After phase rotation and delay compensation the signals are
correlated in pairs in a one bit correlator with 1024 delay channels.
The modular design of the correlator allows up to 16 baselines to be
correlated simultaneously each with 64 delay channels.

　　　　　The observations of OH127.8-0.0 (Booth et al. 1981) were conducted between late 1978 and early 1980 before MERLIN was completed.   3
of the available outstation telescopes were used, each in conjunction
with the Mk IA to give 3 single baseline interferometers (Mk IA-Knockin,
Mk IA-Defford, Mk IA-Mk III).   On each occasion the radio link technique
was used and the data were cross-correlated in the 1024 channel spectrometer to give a 512 point cross-correlation spectrum in the manner
described by Norris & Booth (1981).   From the cross-correlation spectra it
was observed that the sharp feature at -65.7 km s^{-1} (see Fig.2) was
unresolved on all the baselines and so it was used as a reference.   This
was used to correct the phase and amplitude of all the other channels of
the cross correlation spectra in order to remove phase errors introduced
by the differential atmospheric path at the telescopes and for receiver
gain variation.   The 3 data sets were then combined and normalized.   The
new data set was split into velocity intervals and a series of maps
produced by Fourier Inversion and CLEANing (Hogbom 1974), each corresponding to a particular range of velocity in the masing gas.

　　　　　Fig.2   The 1612 MHz OH spectrum of OH127.8-0.0. The bracketed
velocity intervals a,b,c and d represent those ranges of
velocity for which the spatial distribution of the maser
emission has been determined (see Fig. 3.)

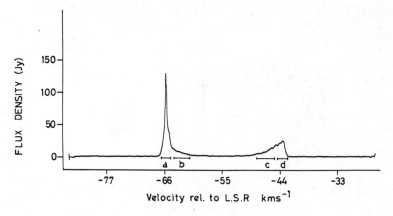

OH104.9+2.4 was observed in the Spring of 1981 using the full
MERLIN procedures and the telescopes Mk II, Mk III, Knockin and Defford.
Data from each telescope, brought back to Jodrell Bank by radio link, were
combined baseline by baseline in the 1024 channel correlator. Since 6
baselines were being processed simultaneously and because of the modular
nature of the correlator, the number of channels available per baseline
was 160. This is equivalent to 80 spectral resolution channels per baseline after Fourier Transformation of the cross correlation function. The
autocorrelation spectrum of OH104.9+2.4 is shown in Fig.4. In this case
no single feature was unresolved on all baselines. Thus a more complex
procedure was adopted to correct the phases of the individual channels.
The feature at -40.3 km s^{-1} was chosen as phase reference and was mapped
using the closure phase technique described by Cornwell & Wilkinson (1981).
When a satisfactory fit to the amplitude and closure-phase data on this
channel had been achieved, the final spatial distribution was Fourier

Fig.3 a-d   Maps of the spatial distribution of the integrated
OH emission of OH127.8-0.0 in the velocity intervals marked a
to d respectively in Fig. 2.   In each map the contour interval
is 5% of the peak integrated emission in that velocity interval
and the lowest contour is ∿10%.

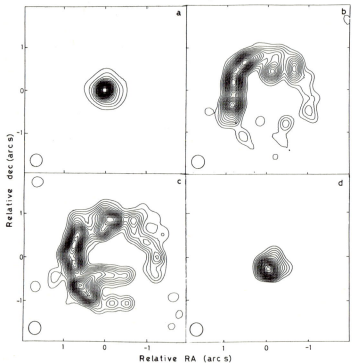

inverted to obtain ϕ_S, the phase due to the source structure. The phases
in the other frequency channels were then corrected using

$$\phi' = \phi - \phi_{ref} + \phi_s - \phi_{BP}$$

where ϕ_{ref} is the observed phase of the reference feature, and ϕ_{BP} is a
small correction term describing the phase response of the instrument.
The corrected phases ϕ' and the observed amplitudes were then processed
using the CLEAN algorithm to produce maps of the emission integrated over
suitable velocity intervals.

The Source Maps

OH127.8-0.0   Maps of the maser emission in four velocity intervals of the
spectra of OH127.8 are shown in Fig. 3. These velocity intervals a,b,c
and d are indicated on the autocorrelation spectra of Fig. 2.

Figure 3a is a map of the distribution of the OH gas with
velocities close to the reference feature and in the range -65 to -66.5
km s^{-1} (region a in Fig.2). It shows that the material within this velocity
interval is concentrated in a small region of angular extent ∿0.4 arc s
and is dominated by the -65.7 km s^{-1} component.   Figure 3d shows the distribution of OH in the corresponding redshifted velocity interval (-43 to
-43.7 km s^{-1}) and is clearly similar in extent although slightly offset in
position.

Fig. 4   The 1612 MHz OH spectrum of OH104.9+2.4. The bracketed
velocity intervals a,b and c are those for which the spatial
distribution is plotted in Fig. 5.

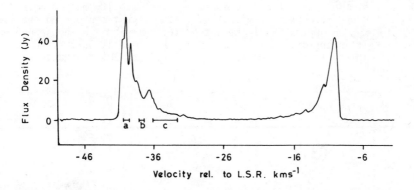

Figures 3b,c show maps of the weaker OH emission in velocity regions closer to the stellar velocity (-62.2 to -64.7 km s^{-1} and -45.2 to -47.5 km s^{-1} respectively). Both have a similar annular form with angular diameter $\sim$2 arc s. The map centres are the same in each case and correspond to the position of the reference feature.

The distributions found are exactly those expected of a uniformly expanding circumstellar shell of OH. The strongest maser components have small angular diameters and are situated close to the line of sight through the centre of the shell along which the gain paths are largest, and the weaker emission delineates those portions of the shell where constant velocity path lengths along the line of sight will be shorter, giving lower maser gain. At the edges of the shell the velocity gradient is such that the line of sight emission is reduced to zero. In this model the radius a(v), at which the line of sight component of the expansion velocity with respect to the star is v, is given by

$$a(v) = R[1-(v/v_e)^2]^{\frac{1}{2}} \qquad (1)$$

where v_e is the expansion velocity and R the total shell radius. In the case of OH127.8-0.0, the expansion velocity (half the interval between OH peaks in Fig.1) is $\sim$11 km s^{-1}. The rings of emission in Fig. 3b,c have mean velocity relative to the star of 8.5 km s^{-1} and angular diameters $\sim$2 arc s. Thus the total shell angular diameter is $\sim$3.1 arc s which corresponds to a linear diameter of 1.5 x 10^{17} cm if we take the kinematic distance to the star to be 3.3 kpc (Bowers et al. 1980).

Fig. 5 Maps of OH104.9+2.4. The maps from left to right are of the integrated OH emission in the velocity ranges marked a, b and c respectively in Fig.3. The lowest contour on the first map is approx 5% of the integrated emission, in the second map it is approximately 10%, and in the third it is 20%. Contour intervals are 5% up to the 50% level and 10% thereafter.

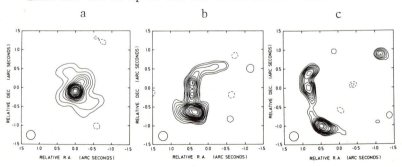

OH104.9+2.4   The spectrum of OH104.9+2.4 is shown in Fig. 4. Although it
has the same overall, twin peaked structure as OH127.8-0.0, this spectrum
is not so smooth and may indicate the presence of non uniform motion in
the shell. This is reflected in the maps of Fig. 5 in which no complete
shells are observed.

The maps of OH104.9+2.4 shown in Fig.5 represent 3 velocity
intervals with negative velocities relative to the star. The feature
mapped in Fig. 5a (-40.3 km s^{-1}) was used as reference and although its
emission is relatively compact, it is not a point source. Fig.5b shows
velocity region b (-38.1 km s^{-1}) of Fig.4. Here the emission delineates
part of a ring with angular diameter $\sim$0.6 arc s. Fig. 5c shows the region
c in Fig.4(-33.0 to -36.6 km s^{-1}) which has an even smaller line of sight
component of velocity. Again part of a ring is apparent but it is clearly
fragmented. Its angular diameter is $\sim$1.2 arc s. Using eqn(1) and an
expansion velocity of 15 km s^{-1} we derive a mean value of 1.3 arc s for
the shell diameter of OH104.9+2.4 from maps 5b and 5c. We take OH104.9+2.4
to be at a distance of 2.5 kpc (Bowers et al. 1980) and so obtain a linear
diameter of 4.9 x 10^{16} cm for the shell.

Discussion

A determination of the mass loss rate from long period variable
stars is of great importance in understanding stellar evolution and the
chemical composition of the interstellar gas. Observations of the circumstellar masers will provide information on the masses involved if the gas
density in the shell can be determined. This involves a determination of
the OH column density through consideration of the brightness temperature
of the maser, its degree of saturation and the efficiency of its pump
mechanism, and an estimate of the abundance of OH relative to H$_2$. In the
case of the sources considered here, individual maser features may be as
small as 30 m arc s. Thus their brightness temperatures are $\sim$10^{9}K. An
efficient 35 μm IR pumping scheme for the masers in circumstellar envelopes
has been developed by Elitzur et al. (1976) and their calculations show
that the 1612 MHz transition is inverted and saturated for OH densities
$N_{OH} \gtrsim 1$ cm^{-3}. Brightness temperatures of $\sim$10^9 are produced for OH
column densities $\sim$10^{17} cm^{-2}, which agrees with our observations. A detailed
model of the circumstellar OH abundance has been developed by Goldreich
and Scoville (1976) who argue that $N_{OH} \gtrsim 1$ cm^{-3}, at a radius of $\sim$10^{16} cm

when the molecular hydrogen number density is $\sim 2 \times 10^4$ cm^{-3}. This value, together with our measured linear diameter gives mass loss rates of 4.3×10^{-5} M$_\odot$ yr^{-1} for OH127.8-0.0 and 2.4×10^{-5} M$_\odot$ yr^{-1} for OH104.9+2.4. Although such mass loss rates are extremely high, they agree with rates derived for a sample of OH/IR stars from measurements of their IR properties, luminosity and dust shell opacities together with the outflow velocities. (Werner, et al.1980). Clearly, stars losing mass at these rates must be short-lived.

A final interesting measurement on the OH/IR stars is that their OH spectra are seen to vary with periods similar to the optical or IR variability of the star. When the OH intensity variations are carefully monitored a phase difference can be observed in the light curve of the blueshifted emission from the nearside of the shell relative to the redshifted emission from the far side. Assuming that the central star is the source of pump radiation, the OH emission from the entire shell should vary in unison and the observed phase lag represents the shell crossing time. Thus we will have a direct measurement of the linear diameter of the shell. By combining this with the angular size determined from interferometry will lead to a direct measurement of the distance of these objects.

References

Booth, R.S., Kus, A.J., Norris, R.P., Porter, N.D. (1981). Nature, 290, 382-4.
Bowers, P.F. (1978). Astron.Astrophys.Suppl., 31, 127-45.
Bowers, P.F., Reid, M.J., Johnston, K.J., Spencer, J.H. & Moran, J.M. (1980). Ap.J., 242, 1088-1101.
Cornwell, T.J. & Wilkinson, P.N. (1981). M.N.R.A.S., 196, 1067-86.
Elitzur, M. (1981). Physical Processes in Red Giants. Eds I. Iben & A. Renzini (D.Reidel).
Elitzur, M., Goldreich, P. & Scoville, N. (1976). Ap.J., 205, 384-96.
Hogbom, J.A. (1974). Astron.Astrophys.Suppl., 15, 417-26.
Goldreich, P. & Scoville, N. (1976). Ap.J., 205, 144-54.
Norris, R.P. & Booth, R.S. (1981). M.N.R.A.S., 195, 213-26.
Reid, M.J., Muhleman, D.O., Moran, J.M., Johnston, K.J. & Schwartz, P.R. (1977). Ap.J., 214, 60-77.
Werner, M.W., Beckwith, S., Gatley, I., Sellgren,K., Berriman, G. & Whiting, D.L. (1980). Astrophys.J. 239, 540-548.

A RANDOM VIEW OF CYGNUS X

Stella Harris
Physics Department, Queen Mary College, Mile End
Road, London E1 4NS

This paper describes the first stage (viz preliminary ^{12}CO and ^{13}CO J = 1 → 0 mapping) in a programme of observations designed to probe the mysteries both of the Cygnus X region and of the statistics of star formation in general. The first section describes the philosophy behind the observations; the second describes the results.

1   Why Random?

Most people involved in the study of star formation, and in particular in looking at questions like how and where stars form within molecular clouds, are only too well aware of the fact that the available statistics are horribly fraught with selection effects. The very first star-forming regions to be known about were those which were made apparent by their optical characteristics, and tended therefore to represent one of two extremes - at the one end the very dense cold 'dark clouds' in which it was unlikely that star formation activity was yet taking place at all; and at the other end the highly evolved 'optical' HII regions, whose configurations with respect to the rest of the molecular cloud would be very different from when the exciting stars were first formed.

Obviously, the advent first of radio and then of infrared and millimetre astronomy has rectified these biases to some extent, but they have nonetheless tended very largely to remain, primarily because the first objects always to be studied with a new technique are usually those known about already through some other means.

Present-day observations of molecular clouds tend largely to fall into two camps - either they are large-beam,

large-area surveys, which are clearly of great value, but
tend not to furnish the fine structure needed for a detailed
study of the locations of star formation; or they are
high-resolution detailed studies of a small area - and
these tend to be exactly those well-known regions mentioned
above, and therefore likely not to represent a statistical
sample.

Hence the primary object of the present observing
programme was to begin to redress this imbalance by
observing in various molecular lines and analysing in detail
a relatively 'random' area of star formation. Inevitably,
with the constraints of telescope time, it had to be a
rather tailored randomness; hence the choice of Cygnus X -
a region known to be sufficiently rich in structure that
one could choose an area fairly well at random without
running a substantial risk of finding an empty field.

The area chosen for study was a 1° square covering:

$$RA \qquad 20^{h} \ 34^{m}.2 - 20^{h} \ 39^{m}.8$$
$$Dec \qquad 41^{o} \qquad - \qquad 42^{o} \ .$$

This was selected for the fact that the region shows no
features of optical interest - apart from the very southern
tip of the Horseshoe Nebula - thereby helping to redress
the imbalance caused by optical selection effects (if
tailoring the randomness yet a little further!) It does,
however, have significant radio continuum structure
(containing parts of the HII regions DR17, DR20, DR22 and
DR23), and also lies close to DR21. Indeed it abuts the
southern boundary of the detailed CO maps made by Dickel et al.
(1978) of the W75/DR21 region. While this latter does not
have the 'randomness' required for the present study - being
well-known as a region singularly rich in star-formation
activity - it is clearly an advantage to be able to combine
these two sets of observations, and obtain detailed CO
information over a relatively large area ($1^{o}.5 \times 1^{o}$),
encompassing regions which could well prove to be very
different in nature.

2 The View

I here present the results of the mapping of the region in the J = 1 → 0 lines of ^{12}CO and ^{13}CO. These data were obtained during two weeks of guest observer time on the University of Austin 5-m telescope during July 1979. For this my sincere thanks are due to the University of Texas, and in particular to Paul Vanden Bout and Bob Loren for all their help and cooperation, and to Glenn White from QMC for coming along and showing me how to use the equipment. I also express my thanks here to Hong-ih Cong of the University of Columbia for the supply of unpublished data (see below).

The whole of the Cygnus X region was in fact mapped by Cong (1977) in ^{12}CO at 8 arcmin spatial resolution using the Columbia 4-ft telescope; he then remapped part of the area at 2.5 arcmin resolution using the Texas dish. The 4-ft results, whilst presenting a useful overall picture, do not give either the spatial or the spectral resolution required for the present study; but the 5-m results - with a 2.5 arcmin beamwidth and a velocity resolution of 0.65 km/s were adequate at least for obtaining the basic structure. Cong kindly supplied me with his 5-m data pertaining to the square degree in question, so that I was able simply to supplement this by observing the areas which he hadn't covered.

The area was surveyed first of all with a coarse grid spacing of 5 arcmin, and then the more structured regions filled in with a finer 2.5 arcmin grid. Excluding the parts already done by Cong, this resulted in a total of 490 ^{12}CO spectra. These spectra were obtained by position-switching with a beamthrow of 140 arcmin, and typical integration times of about 5 minutes. They were calibrated against DR21, which was observed every hour.

There are two essential problems encountered in analysing this region. The first is actually untangling the different velocity components; and the second is then deciding how, if at all, they are related to each other.

In the ensuing analysis the velocities are divided into
four main groups, centred at around -3, +5, +8 and +15 km/sec.
This fits well with other analyses of the region in general,
though the situation is not completely clear-cut, as
features exist at almost all intermediate velocities (an
interesting point which I'll return to), and there are also
occasions when one finds more than four individual components
on any one map. Figs 1 to 4 show the ^{12}CO maps in these
four components. It can be seen that the +5 km/s
component seems to be a fairly ubiquitous and amorphous
structure - a diffuse cloud (or probably a combination of
diffuse clouds) pervading the whole area. The inter-relation
of the various components will be discussed below.

The ^{12}CO hotspot regions were subsequently
mapped in ^{13}CO (about 150 spectra in all); these maps are
also shown in Figs 1-4, except for the 5 km/sec component,
for which there was only one point at which the ^{13}CO
temperature exceeded 3 K. Individual features of interest
are mentioned in the captions. Of particular interest is
the evidence for at least one area of ^{12}CO self-absorption,
in the 8 km/sec feature, demonstrated in Fig. 3. Here
the ^{13}CO hotspot appears to extend northwards into a dip
in the ^{12}CO temperature. However, closer inspection of
the spectra (Fig. 5) shows that this apparent dip in the
^{12}CO is almost certainly produced by self-absorption.
Individual spectra are also shown (Fig. 6) for the area
around the main 15 km/sec hotspot to the northwest of the
region, to illustrate the very sharp edges of this feature
and the broadening of both lines towards the southeast. The
product $T.\Delta v(^{13}CO)$ yields a direct measure of ^{13}CO column
density (as given, for example, by Evans et al. 1977). The
areas of enhanced ^{13}CO temperature and density were found
to be broadly coincident, the latter attaining peak values
of $n_{13}L \sim 4 \times 10^{16}$ cm^{-2}.

Returning to the velocity structure, Cong in
his thesis argues in favour of the 8 km/s component
representing a foreground cloud, with all the others being

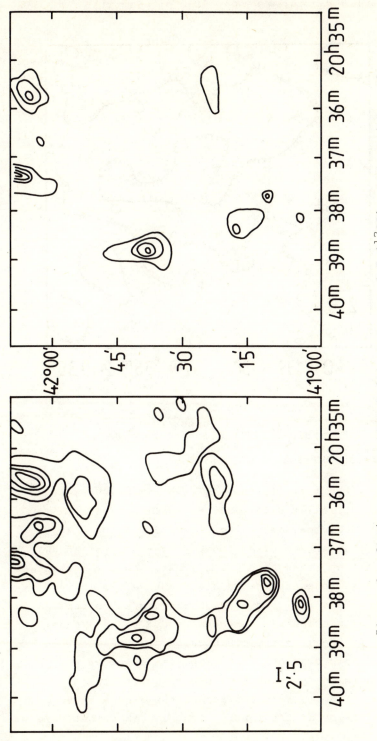

Figure 1: -3km/sec component. Left: contours of $T_A(^{12}CO)$; contour interval is 5K; lowest contour is 5K. Right: contours of $T_A(^{13}CO)$; contour interval is 2K; lowest contour is 3K. The northeasternmost feature is DR21.

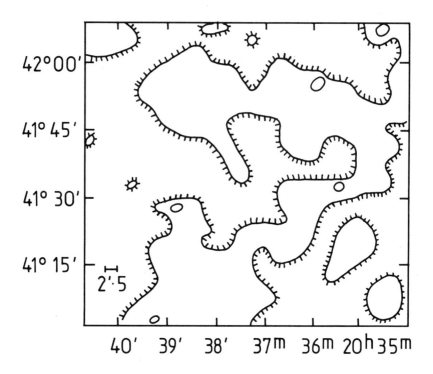

Figure 2: +5km/sec component, contours of $T_A(^{12}CO)$; contour interval is 5K; lowest contour is 5K. Hatching indicates direction of decreasing brightness. As discussed in the text, $T_A(^{13}CO)$ exceeds 3K at only one point.

at essentially the same (further) distance. However, this is refuted by Dickel et al. (1978) in their analysis of the W75 region - where they detect both the -3 and the 8 km/s components. They argue quite strongly in favour of an interaction between these two clouds, based on the facts that the most intense hotspots in the different velocities tend always to be closely adjacent but never coincident, indicating possible density enhancement due to collision; and the fact that all intermediate velocities seem to be present to some degree - again implying an actual interaction between the clouds. They also cite the presence of a velocity gradient across the 8 km/s cloud as further evidence of its interaction with the other.

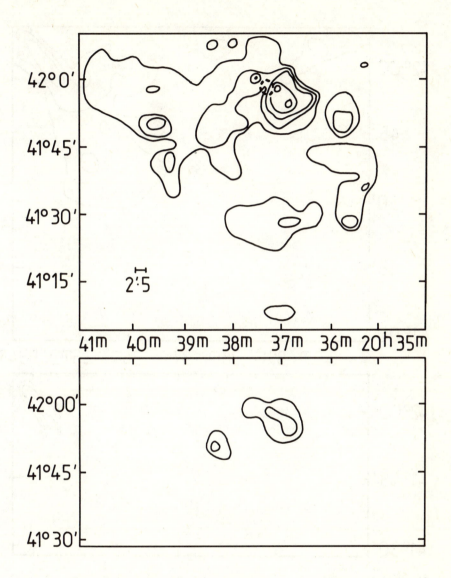

Figure 3: +8km/sec component. Above: contours of
$T_A(^{12}CO)$; contour interval is 5K; lowest contour is
5K. Probable region of ^{12}CO self-absorption is
indicated with a dashed ring (see also Fig. 5).
Below: 2K and 5K contours of $T_A(^{13}CO)$.

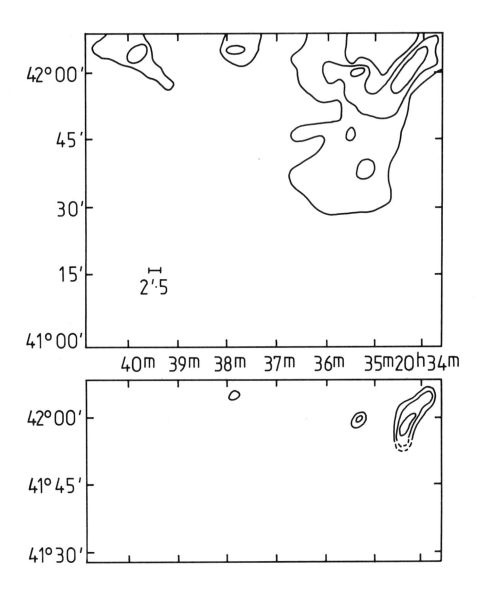

Figure 4: +15km/sec component. Above: contours of $T_A(^{12}CO)$; contour interval is 5K; lowest contour is 5K. Below: 3K and 5K contours of $T_A(^{13}CO)$. Most interesting feature is the narrow and steep ridge to the northwest (see also Fig. 6).

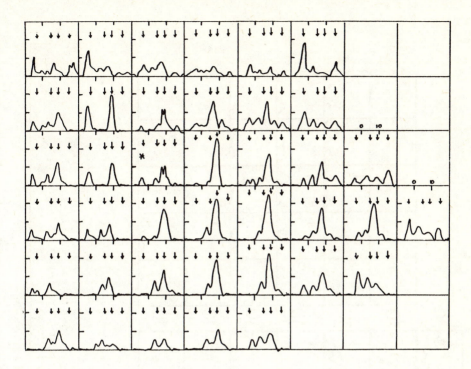

Figure 5: Sketches of ^{12}CO spectra in the region of
suspected ^{12}CO self-absorption at 8km/sec.
Horizontal scale represents km/sec (tick marks
at 0 and 10km/sec, and the four arrows in each
sketch representing -3, +5, +8 and +15 km/sec);
vertical scale represents T_A(K) (tick marks at
10K and 20K). The asterisk denotes the position
of ^{13}CO maximum antenna temperature.

Figure 7 shows a combined map of the ^{12}CO structure at
temperatures >10 K for all the velocity components. The
same point made by Dickel et al. for W75 is seen to pertain
over the whole of this region - that the hotspots in the
different components tend always to lie close together but
not to coincide. As mentioned earlier, the presence of
all intermediate velocities as well as the main ones is also
seen over the whole of this area, thereby providing some
evidence that all of these clouds are in some way interacting.
It is also interesting to note that a velocity gradient is
also observed across the main 8 km/s component detected
in this region - which lies to the south of DR21.

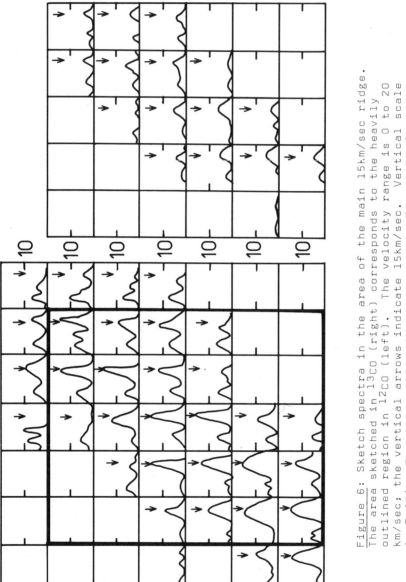

Figure 6: Sketch spectra in the area of the main 15km/sec ridge. The area sketched in 13CO (right) corresponds to the heavily outlined region in 12CO (left). The velocity range is 0 to 20 km/sec; the vertical arrows indicate 15km/sec. Vertical scale is TA(K). The interesting features are the (spatially) rapid appearance and disappearance of this component (in comparison to the fairly uniform weaker presence of the 5km/sec feature) and the broadening of the line to the southeast.

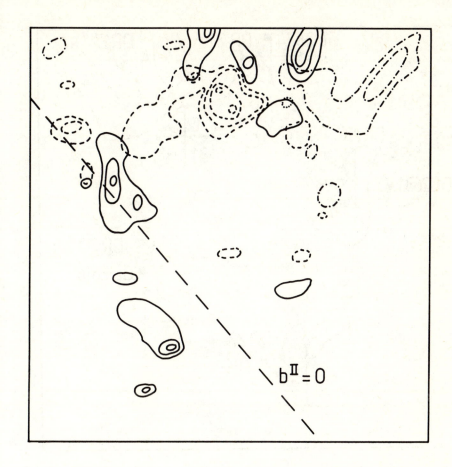

Figure 7: Composite map showing ^{12}CO hotspots (>10K) at all velocities. The components are -3km/sec (——), +5km/sec (·· ·· ·· ··), +8km/sec (- - - - -) and +15km/sec (-·-·-·-).

        Fig. 8 shows again the outlines (10K contour) of the ^{12}CO hotspots, together with the main radio HII regions (taken from the map by Wendker, 1970), and the known infrared sources in the region - these being taken from the AFGL catalogue (4,11,20 (Price & Walker 1976) and from the 12' beam survey by Campbell et al. (1980) at 82 and 92 μm.

        The main features of interest, and particularly deserving of further observation, are:

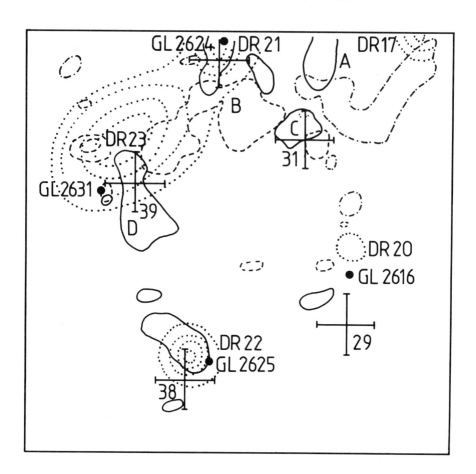

<u>Figure 8</u>: Map showing outlines of ^{12}CO hotspots
(10K contour), with known infrared sources and radio
HII regions. Infrared sources are from AFGL
(Price & Walker 1976) shown as dots and from
Campbell et al. 1980 (crosses); radio sources
(dotted contours) are from Wendker 1970. The main
regions of interest, labelled A, B, C, D, are
discussed in the text.

A. Represents a strong ^{12}CO and ^{13}CO hotspot, not associated
   with any known HII region or infrared structure,

B. ^{12}CO self-absorption feature,

C. Has no close association with any known radio feature, but
   a far-infrared source and a meeting point of all the
   molecular velocity components,

D. Represents a strong ^{12}CO and ^{13}CO hotspot, with the ^{13}CO
   peak possibly displaced slightly to the south of the
   ^{12}CO peak (but no sign of self-absorption in the ^{12}CO
   spectrum).

These will be investigated further in future stages of the
observing programme.

References

Campbell, M. F., Hoffmann, W. F., Thronson, H. A. & Harvey,
         P. M., (1980), Astrophys. J., __238__, 122.
Cong, H. L. (1977), Ph.D. thesis, Columbia University.
Dickel, J. R., Dickel, H. R. & Wilson, W. J. (1978),
         Astrophys. J., __223__, 840.
Evans, N. J., Blair, G. N. & Beckwith, S. (1977),
         Astrophys. J., __217__, 448.
Price, S. D. & Walker, R. G.,(1976), AFGL-TR-76-0208
Wendker, H. J.,(1970),Astron. & Astrophys. __4__, 378.

THE M(17)SW MOLECULAR SOURCE: INTERNALLY OR
EXTERNALLY HEATED?

Marcus P. Chown, John E. Beckman, Nigel J. Cronin
and Glenn J. White
Department of Physics, Queen Mary College,
Mile End Road, London E1 4NS. England.

Abstract. From a measurement of the $J = 3 \rightarrow 2$
rotational emission line of hydrogen isocyanide,
and a literature value for that $J = 1 \rightarrow 0$
line (Snell and Wootten, 1979) we have derived
the excitation parameters for HNC in the direction
of the molecular peak as determined by CO. Kinetic
temperatures for the optically thick CO and ^{13}CO
transitions are ∿50 K, while that pertaining to
HNC line formation is between 20 K and 5 K. The
inference that the central temperature of the
cloud is lower than that of its surface, supports
models in which the cloud is heated not from
within, but by the O-B association to the north-
west.

1   Introduction

The immense scale of the molecular clouds close
to the M17 HII region was revealed first by the CO mapping
of Lada, et al. (1974), and then by the even more extended
survey of Elmegreen and Lada (1976). The M17(SW) cloud is the
brightest knot in the ∿85 pc complex, and the second
brightest molecular source in the sky. It is juxtaposed
with an HII region, the Omega nebula, which is itself excited
by an OB cluster. The HII region takes the form of a
'blister' at the north-west edge of the cloud. A map of
the cloud in ^{12}CO, (Lada 1976) together with our $J = 3 \rightarrow 2$
map of HNC is shown in Fig. 1. The CO peak is at
$\alpha_{1950} = 18^h17^m26^s$, $\delta_{1950} = -16^o15'$, shown in Fig. 1.
and our HNC peak is 2 arc minutes north of this, in almost
exact coincidence with the recent C_2H peak of Wootten et al.
(1980). The majority of excited regions in molecular
clouds are heated from within by protostars or newly

formed stars, and it was believed that the KW object
(Kleinmann and Wright, 1973) might fill this rôle for the
M17(SW) cloud. However, the energy output of this 10 μm
source falls short of the power required by some two
orders of magnitude. More recently a model by Icke et al.
(1980) which accounts for far infrared and radio continuum
observations of the cloud by Gatley et al. (1979) and Wilson
et al. (1979), suggests that the whole cloud may be
heated by the OB cluster. The displacement of the HNC
and C_2H peaks by ∿1.5 pc from the CO peak, and the intensity
data described below, lend significant support to this
model, and point the way to detailed radiative transfer
pictures of the whole cloud.

Figure 1 Maps of the HNC antenna temperature for
M17(SW) in the J = 3 → 2 line (dashed line)
contours at 0.5 K, 1.0 K, 1.5 K, 2.0 K and
2.5 K; and ^{12}CO antenna temperature in J = 1 → 0
line (solid line), contours at 15 K, 20 K, 30 K,
40 K and 50 K, (Lada, 1976).

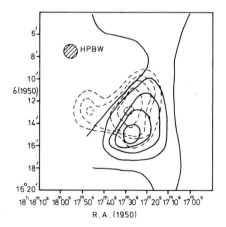

2   Temperature of formation of HNC

Figure 2 shows a spectrum of the cloud in the
HNC J = 3 → 2 transition, observed at the UKIRT 2.8 m telescope
towards the CO intensity peak (N.B. not towards the peak of
HNC emission, whose peak antenna temperature was more than
twice the value of 1.2 K shown in Figure 2).   In comparing
this line with the J = 1 → 0 line of HNC measured on the
same direction by Snell and Wootten (1979) we must take
into account the effect of hyperfine structure.   The
observed J = 3 → 2 line is a blend of hyperfine components
whereas for the J = 1→ 0 line these components are well
separated, so that the F = 2 → 1 line is observed in isolation.
Thus to make a proper intensity comparison, we reduce our
corrected antenna temperature T_A^* to 0.74 ($\bar{+}$0.15) K; the
corresponding J = 1 → 0 temperature is 2.24 ($\bar{+}$0.22) K.

Using an isothermal, constant density model we can
rapidly obtain an excitation temperature for the HNC
molecules, on the basis of thermodynamic (Boltzmann)
equilibrium, and the Rayleigh-Jeans approximation.   For a
corrected antenna temperature T_A^*, the excitation temperature
T_{ex} for a line of optical depth τ is given by:

Figure 2   Velocity spectrum of the HNC J = 3 → 2
transition at 271.981 GHz, towards the CO intensity
peak of M17(SW).

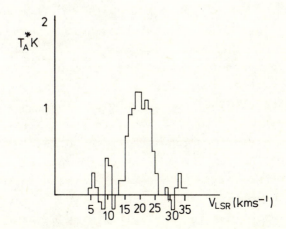

$$T_A^* = (1.22)^2 \frac{h\nu}{k} \left| (e^{(h\nu/kT_{ex})} - 1)^{-1} - (e^{(h\nu/kT_c)} - 1)^{-1} \right| (1 - e^{-\tau}) \quad (1)$$

where T_c is the background continuum temperature of ~2.9 K.
This can be solved for T_{ex}, given the measured values of
T_A^* for two lines and assuming a common excitation temperature.
If the number of molecules in successive rotational
levels is linked by the Boltzmann factor $e^{-h\nu/kT_{ex}}$ the
equations may be solved for T_{ex}, $\tau_{3 \to 2}$ and $\tau_{1 \to 0}$, giving
T_{ex} = 4.8 K, $\tau_{1 \to 0}$ = 1.12, $\tau_{3 \to 2}$ = 0.85. The range of
abundances X_{HNC} is set between 4.3×10^{-12} and 5×10^{-11},
where $X_{HNC} = |HNC| : |H_2|$. This range is derived from the
work of White et al. (1982) for HCN, and from the HNC paper
by Snell and Wootten (1979) for this source. Given this
abundance range, our value for $\tau_{3 \to 2}$, and the radius of
the cloud as 2 pc, we can derive a density range for the
cloud of $1.4 \times 10^5 < n_{H_2} < 1.2 \times 10^6$ for the hydrogen
molecule number density in cm^{-3}. Application of a
simplified low velocity gradient model to a cloud with
these values for n_{H_2} and T_{ex} gives a kinetic temperature
range of 20K > T_{KIN} > 5K. We can avoid the step of relying
too heavily on estimates of X_{HNC} by using literature values
of n_{H_2}, obtained from other molecular species. The paper
by White et al. (1982) gives $n_{H_2} \sim 5 \times 10^5$ cm^{-3} from an
LVG model based on the HCN J = 4 → 3 line, and Wootten et al.
(1980) find $n_{H_2} \sim 7 \times 10^5$ from the J = 1 → 0 line of
HCO^+ and the 2(1,2) → 1(1,1) line of H_2CO. A best kinetic
temperature from our HNC data is then $T_K \sim 10$ ($\bar{+}3$) K.
Our range of values for T_{KIN} is weighted towards the
densest part of the cloud, since excitation falls off with
decreasing density. Thus we can conclude that the CO
kinetic temperature of 50 K pertains on those parts of the
cloud surface receiving UV photons from the OB cluster,
while the dust is shielding the denser central region.
This view is strengthened by the displacement of the HNC
peak towards the cluster shown in Fig. 1, but a more
comprehensive model dealing explicitly with radiative

transfer through the dust is in preparation with D. Walker (see also this volume).

3 Density distribution within the cloud

From the temperature isophotes of Fig. 1 and the assumption that the emission is not optically thick, we can, using the procedures outlined in section 2, derive a map of the optical depth $\tau_{3 \rightarrow 2}$ of the HNC J = 3 → 2 transition, which is presented in Fig. 3. A density distribution of gaussian form

$$n = n_o \, e^{-(\frac{r}{r_a})^2} \qquad (2)$$

where n_o is the density at the centre, and r_a is the 1/e-fold radius will give a projected optical depth variation of gaussian form also, viz.:

$$\tau = \tau_o \, e^{-(\frac{p}{r_a})^2} \qquad (3)$$

A best fit of equation (3) to the data in Fig. 3 yields 2.2 pc. for r_o, and assuming a central density n_o of 5×10^5 cm^{-3}, we have an average cloud density <n> of 3.3×10^5 cm^{-3} out to this equivalent radius. This

Figure 3 Optical depth contours for the HNC J = 3 → 2 transition in the M17(SW) molecular cloud; contour values for $\tau_{3 \rightarrow 2}$ are

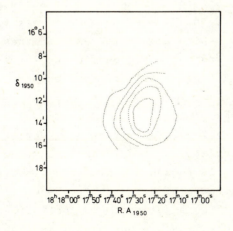

relatively slow fall-off in density is consistent with a
cloud which, although massive ($M_{cloud} \sim 5 \times 10^5 M$) is not compressed
towards the core. The wide HNC line of Fig. 2 ($\Delta v_{FWHM} \sim 15$ km s^{-1})
implies a highly turbulent core, resistant to collapse. The
size of the cloud does, however, imply a great optical depth
in the visible due to dust. Taking our values for $<n>$ and r_a,
together with literature values for $N(^{13}CO)/N(H_2)$
(Tucker et al. 1976) and for $N(^{13}CO)/A_v$ (Beckman and
Moorwood 1979) where $N(^{13}CO)$ is the number density of
^{13}CO, $N(H_2)$ the same for hydrogen molecules, and A_v
the visual extinction due to dust, we obtain

$$A_v = 3.13 \times 10^3, \text{ i.e. } \tau_v = 3.4 \times 10^3 \qquad (4)$$

for the visual extinction and optical depth, respectively.
Although there is no guarantee that the standard literature
ratios apply to any particular cloud, it is clear that the
visual optical depth is $\gtrsim 10^3$. This compares with a value
of ~ 100 obtained by Gatley et al. (1979) at the CO
peak by extrapolating from the measured optical depth
of dust at 50 µm.

    With our value of optical depth, we can
assert, at least qualitatively, using the models of
Leung (1978) that temperature gradients across the M17(SW)
cloud, from the OB association in the north west, to the
HNC molecular peak, of order 100 K could be sustained
by the dust blanket. The UV stellar photons are trapped
by the outer layers, which are heated and can then
re-radiate into the rest of the cloud in the infrared.
The cool-centre is the net result of this eccentric geometry.

    Conclusions
    The HNC J = 3 → 2 transition has four properties
which make it particularly suited to exploring the M17(SW)
molecular cloud. It is reasonably pressure sensitive, and not
optically thick. It has a high enough frequency (272 GHz)
to yield reasonably high angular resolution (~ 1 arcmin)
with the 3.8 metre UKIRT telescope; and it is probable

that the formation conditions favour its concentration in
relatively low temperature regions of molecular clouds. We
have used our data to derive the basic physical parameters
for the cloud density distribution, central temperature,
and temperature gradient. These are all consistent
with an externally heated, cold-centred cloud, and
numerically with the presence of the M17 HII region/OB
complex to the north west.

References

Beckman, J. E. and Moorwood, A. F. M., 1979. Reports on
          Progress in Physics, 42, 87.
Elmegreen, B. G. and Lada, C. J., 1976, Astron. J. 81, 1089.
Elmegreen, G., Lada, C. J. and Dickinson, D. F., 1979.
          Astrophys. J., 230, 415.
Gatley, I., Becklin, E. E., Sellgren, K. and Werner, M. W.
          1979. Astrophys. J., 233, 575.
Icke, V. , Gatley, I. and Israel, F., 1980. Astrophys. J.
          236, 465.
Kleinmann, D. E. and Wright, E. L., 1973. Astrophys. J.,
          185, L131.
Lada, C. J., Dickinson, D. F. and Penfield, H., 1974.
          Astrophys. J. 189, L35.
Lada, C. J., 1976. Astrophys. J., Suppl. Ser., 32, 603.
Leung, C. M., 1978. Astrophys. J. 222, 140.
Snell, R. L. and Wootten, A., 1979. Astrophys. J. 228, 748.
Tucker, K. D., Dickman, R. L., Encrenaz, P. J. and Kutner,
          M. L., 1976. Ap. J. 210, 679.
White, G. J., Beckman, J. E., Cronin, N. J. and Phillips,
          J. P., 1982. Mon. Not. Roy. astr. Soc.,199,375.
Wilson, T. L., Fazio, G. G., Jaffe, D., Kleinmann, D.,
          Wright, E. L. and Low, F. J., 1979. Astr. Astrophys.
          76, 86.
Wootten, A., Snell, R. and Evans, N. J. II,1980. Ap. J.
          240, 532.

OBSERVATIONS OF ROTATIONAL TRANSITIONS OF CH_3OH AND NH_2 NEAR 1.2 MM

Th. de Graauw[*] and S. Lidholm[*], Astronomy Division, Space Science Dept. of ESA, Noordwijk, the Netherlands – W. Boland, Astronomical Institute, University of Amsterdam, the Netherlands – T.J. Lee[*], United Kingdom Infrared Telescope Unit, HILO, Hawaii, USA – C. de Vries[*], Astronomical Institute, University of Utrecht, the Netherlands.

[*] Visiting astronomer of the United Kingdom Infrared Telescope and the Infrared Telescope Facility which is operated by the University of Hawaii under contract of NASA.

Abstract. Sixteen new methanol lines have been detected at frequencies near 242 GHz towards the direction of OMC-1. They correspond to the $J = 5 \to 4$, sk = 0 transitions in the A and E stack of CH_3OH. Also towards OMC-1 we observed a broad emission feature, coinciding in frequency with $J = 7/2 \to 5/2$ components of the $3_{13} \to 2_{20}$ rotational transition of NH_2. The observed antenna temperatures of the methanol lines can be reproduced with a thermal excitation model using one temperature $T_{exc} = 90K$ and a column density of $N(CH_3OH) = 5 \times 10^{16}$ cm^{-2}, assuming a 30" diameter methanol source.

Introduction

Methanol (CH_3OH) has already been observed in about 35 transitions towards the Orion Molecular Cloud (Barrett et al. 1971; Kutner et al. 1973; Lovas et al. 1976; Gottlieb et al. 1979, Jennings and Fox, 1979).

Most lines seem to be consistent with thermal excitation at one temperature, $T_{exc} = 90K$. However, some transitions have intensities that vary in time and have narrow, multiple velocity components (Barrett et al. 1975; Hills et al. 1975) characteristic for maser emission. Matsakis et al. (1980) have mapped OMC-1 in two methanol transitions using the Hat Creek interferometer. They resolved the source into 10 components, with different velocities, seperated by at most 30".

At a frequency near 242 GHz we detected 16 new methanol lines produced by the $J = 5 \to 4$, sk = o transitions of CH_3OH. These lines originate from levels with upper level energies ranging from 26 to 87cm^{-1} above the ground state.

Near 241.5 GHz we detected a broad (150 km s^{-1}) emission feature coinciding with a series of fine structure components of the $3_{13} \to 2_{20}$ rotational transition of the radical NH_2, which frequencies have recently been determined by laboratory experiments (Charo et al., 1981). NH_2 plays an important role in the general scheme of formation and

decomposition of larger molecules in dense interstellar clouds. It is a
light asymmetric rotor, and its electric dipole transitions between the
low lying energy levels lie at submm and near mm wavelengths. It has
been detected in cometary nuclei but not in interstellar space (Giguere
and Clark, 1975). The expected abundance of the NH_2 radical in dense
clouds from gas phase reactions is highly uncertain. The main reason is
that different measurements of the reaction rate of $NH_3^+ + H_2 \rightarrow NH_4^+ + H$
give different values (Fehsenfeld et al., 1975; Smith and Adams, 1980).
Decomposition of NH_3^+ and NH_4^+ lead to formation of NH_2 and NH_3 and therefore detection of NH_2 will help to clear the NH_3 chemistry picture.

Observations

The observations were carried out in October 1980 with the
3.8m diameter U.K. Infrared telescope and NASA's 3m Infrared Telescope
Facility, both located on top of Mauna Kea, Hawaii. The receiver used
has been described elsewhere (Lidholm and de Graauw, 1979). It had room
temperature Schottky diode waveguide mixers and a carcinotron local
oscillator. The system noise temperature amounted 4000 K SSB. The

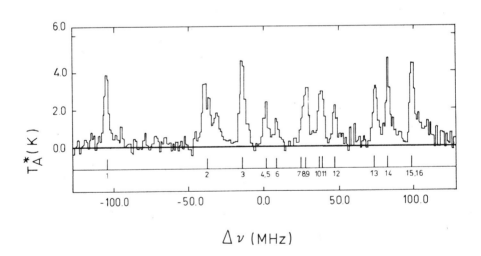

Figure 1. Methanol lines observed towards Ori A.

antenna beam widths and the efficiencies of the antenna-receiver combinations were determined by scans across the Moon and Jupiter and we
obtained 90"-55% and 110"-50% for UKIRT and IRTF respectively. Two filterbanks (1 MHz and .25 MHz resolution) were used in parallel. The 256 x
1 MHz bank provided a velocity span of about 300 km sec^{-1}. The atmospheric transmission obtained from sky dips was between 80 and 90%.
Both telescopes provide stable baselines when used in a position switching
observing mode with 30' throw at 100 seconds intervals. IRTF's small
(chopping) secondary is probably responsible for slightly better baselines.

Observations were made mainly towards the K-L nebula
(α(1950) = $5^h32^m47^s$, δ(1950) = $-5^o24'21"$) and 2' south.

Results and discussion
CH_3OH

Identification of the methanol lines was based on the frequencies given by Lees (1973) and Pickett (1980). Figure 1 shows the lines
observed towards OMC-1 with UKIRT. The vertical bars indicate the position of the 16 transitions which are labeled as in the energy level diagram shown in figure 2 and as in table 1. The antenna temperatures are
corrected for atmospheric transmission and telescope losses. The line-
widths are approximately 4km s^{-1} (FWHM) in agreement with other methanol
transitions observed towards the K-L nebula. The feature at -31 MHz is
an unidentified line (241.774 MHz).

Table 1 summarizes the observed and calculated peak brightness temperatures T_B together with the relevant data used in the calculations. The spontaneous radiative transition probabilities A_u have been
taken or calculated from Lees (1973) and Lees et al. (1973).

The brightness temperature is derived from the peak antenna
temperature, T_A^*, using the relation $T_B = \{\Theta_A/\Theta_s\}^2 T_A^*$, where Θ_A and Θ_s are
the angular diameters of the beam and source, respectively. Most of the
maser spots observed by Matsakis et al. (1980) are located within a radius
of 15" from the Kleinmann-Low object. Assuming that the non-maser lines
also originate from the region in between the maser spots we get a source
diameter $\Theta_s \sim 30"$. This size agrees with the upper limit estimated by
Hills et al. (1975) and is consistent with the ratio in antenna temperatures we obtain using both telescopes.

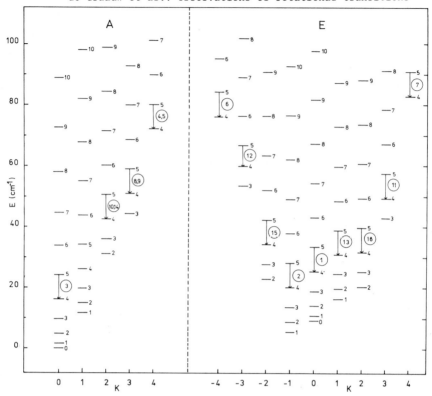

Fig. 2. Energy level diagram of CH3OH A and E species. The detections discussed in this paper are given by arrows and their labels refer to Table 1.

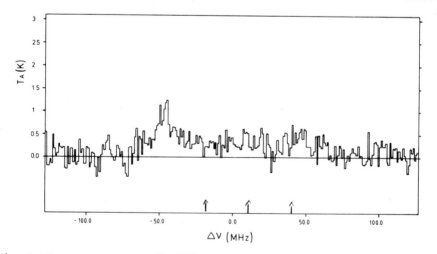

Fig. 3. Spectrum taken at 241.572 MHz showing a broad emission plateau. Arrows indicate the three strongest fine structure components of $NH_2 (3 \rightarrow 2)$.

Table 1: Properties of the CH_3OH transitions observed towards OMC-1

Number	Transition	ν_{ul} (GHz)	A_{ul} (in 10^{-5} s^{-1})	T_B (K) obs.	T_B (K) calc.
1	$5_0 \rightarrow 4_0$ E	241.7002	5.85	34	34
2	$5_{-1} \rightarrow 4_{-1}$ E	241.7672	5.62	30	35
3	$5_0 \rightarrow 4_0$ A$^+$	241.7914	5.86	40	45
4	$5_4 \rightarrow 4_4$ A$^+$	241.8065	2.11	} 17	16
5	$5_4 \rightarrow 4_4$ A$^-$	241.8065	2.11		
6	$5_{-4} \rightarrow 4_{-4}$ E	241.8133	2.11	12	7
7	$5_4 \rightarrow 4_4$ E	241.8296	2.11	blend	6
8	$5_3 \rightarrow 4_3$ A$^+$	241.8330	3.75	} 27	35
9	$5_3 \rightarrow 4_3$ A$^-$	241.8330	3.75		
10	$5_2 \rightarrow 4_2$ A$^-$	241.8423	4.92	25	28
11	$5_3 \rightarrow 4_3$ E	241.8436	3.75	blend	16
12	$5_{-3} \rightarrow 4_{-3}$ E	241.8524	3.75	17	14
13	$5_1 \rightarrow 4_1$ E	241.8791	5.63	28	30
14	$5_2 \rightarrow 4_2$ A$^+$	241.8877	4.92	32	28
15	$5_{-2} \rightarrow 4_{-2}$ E	241.9041	4.93	} 35	45
16	$5_2 \rightarrow 4_2$ E	241.9044	4.93		

We predict brightness temperatures by supposing that the CH_3OH energy levels are populated according to one excitation temperature T_{exc}. Then the peak brightness temperature, T_B , is calculated to be

$$T_B = T_{exc} \{1 - \exp(\tau_{lu})\} \tag{1}$$

where the line centre optical depth τ_{lu} is given by

$$\tau_{lu} = \frac{A_{ul} c^3}{4 \pi \nu_{ul}^3} \left(\frac{\ln 2}{\pi}\right)^{\frac{1}{2}} \frac{N_1}{\Delta V} \frac{g_u}{g_1} \{1 - \exp\left(\frac{-h \nu_{ul}}{k T_{exc}}\right)\} \tag{2}$$

The quantities g_u and g_1 are the statistical weights of the upper and lower level, respectively, ΔV is the full line width at half maximum and N_1 is the column density of molecules in level 1. N_1 is related to the total CH_3OH column density according to

$$N_1 = \frac{g_1 \, N_{total}}{2 \, Q_i} \quad \exp \left(\frac{-h \, E_1}{k \, T_{exc}} \right) \tag{3}$$

where E_1 is the energy of level 1 and Q_i is the rotational partition function for the methanol A or E species. Q_E is taken from Lees (1973) while Q_A is calculated using Lees' prescription and including levels up to J=12 and k=4. The factor 2 in Eq. (3) is included because the CH_3OH A and E species are strictly separated with approximately equal numbers of molecules in each species.

The brightness temperatures T_B(calc.) presented in Table 1, are obtained for T_{exc} = 90 K, $N(CH_3OH)$ = 5 x 10^{16} cm^{-2} and ΔV = 4 km s^{-1}. Within the errors of the observed brightness temperatures (± 3.5 K) we find a good agreement between the calculated and observed line intensities. The estimated uncertainties are T_{exc} = 90 ± 20 K and $N(CH_3OH)$ = 5(-3, +2) x 10^{16} cm^{-2}.

NH2

Observations in a frequency range of 250 MHz wide around 241.572 GHz rendered a broad emission like feature with an emission line super imposed at 241.616 GHz as shown in figure 3. In order to establish the nature of this feature and to eliminate the possibility of instrumental origin, we shifted the BWO local oscillator first by 11 MHz to establish in which sideband the emission was likely to be detected, and subsequently moved up in frequency by twice the IF value. In both cases we again found the broad emission but after the second shift its appearance was mirrored, as indeed it should be, since the software was not corrected for sideband. After these observations we were informed by Dr. de Lucia and collaborators about the location of the 3-2 transitions of NH_2. The position of the three strongest fine structure transitions are given in figure 3, see de Lucia in this conference proceedings. We are therefore inclined to believe that the broad emission arises from NH_2 in the high velocity source in Orion-A. The spike feature has been identified with an SO_2 transition.

We have calculated the NH_2 column density assuming T_{ex} = 70 K, a source diameter of 30" and a velocity spread ΔV (FWHM) of 50 km s^{-1}, using formulae (1), (2) and (3). In the calculations of the transition probabilities we used a value for the dipole moment (1.9 Debye) calculated

by Higuchi (1956) and we have taken into account the relative intensity
distribution as given by Charo et al. (1981). For a corrected antenna
temperature $T_A^* = .5$ K we obtain $N(NH_2) \sim 2 \times 10^{17}$ cm^{-2}. A comparison with
NH_3 would be valuable, but recent VLA observations of NH_3 in several of
its transitions show a very clumpy structure which varies for the differ-
ing transitions. Wilson et al. calculated column densities of the (1,1)
and (2,2) lines of NH_3 with $T_{ex} = 70$ K using Effelsberg observations, and
obtained 1.4×10^{16} cm^{-2} and 1.1×10^{16} cm^{-2}. In comparison, the column
density for NH_2 appears to be relatively high, and requires a large NH_2
production rate compared with NH_3. Further observations of the fine
structure transitions at 229 GHz are being made to verify the detection
of NH_2.

Acknowledgements

We thank the staff of UKIRT and IRTF for their continuous
technical support during the observation period.

References

Barrett, A.H., Schwartz, P.R., Waters, J.W., Astrophys. J. Letters 168,
      L10 , 1971.
Barrett, A.H., Ho, P., Martin, R.N., Astrophys. J. Letters 198, L119,
      1975.
Charo, A., Sastry, K.V.L.N., Herbst, E. and De Lucia, F., Astrophys.
      J. 244, L111, 1981.
Cohen, E.A., Schafer, M.M., J. of Molec. Spectr. 89, 1981.
Fehsenfeld, see Charo, 1975.
Giguere, P.T. and Clark, F.O., P. J. 198, 761, 1975.
Gottlieb, C.A., Ball, J.A., Gottlieb, E.W., Dickinson, D.F., Astrophys. J.
      227, 422, 1979.
Higuchi, J. Chem. Phys. 24, 535, 1956.
Hills, R., Pankonin, V., Landecker, T.L., Astron. Astrophys. 39, 149,
      1975.
Jennings, D.E., Fox, K., Astrophys. J. 227, 433, 1979.
Kutner, M.L., Thaddeus, P., Penzias, A.A., Wilson, R.W., Jefferts, K.B.,
      Astrophys. J. Letters 183, L27, 1973.
Lees, R.M. Astrophys. J. 184, 763, 1973.
Lees, R.M., Lovas, F.J., Kirchhoff, W.H., Johnson, D.R., J. Phys. Chem.
      Ref. Data 2, 205, 1973.
Lidholm, S., de Graauw, Th., in Fourth International Conference on Infra-
      red and Millimeter waves and their Applications, S. Perkowitz
      (ed.), Appendix p.38, IEEE cat. no. 79, CH 1384-7, MIT, 1979.
Lovas, F.J., Johnson, D.R., Buhl, D., Snyder, L.E., Astrophys. J. 209,
      770, 1976.
Matsakis, D.N., Cheung, A.C., Wright, M.C.H., Askne, J.I.H., Townes, C.H.,
      Welch, W.J., Astrophys. J. 236, 481, 1980.
Pickett, H.M., Cohen, E.A., Phillips, T.G. Astrophys. J. Letters
      236, L 43, 1980
Smith and Adams, see Charo, 1981.

HIGH VELOCITY FLOWS AND MOLECULAR JETS

Charles J. Lada
Steward Observatory, University of Arizona, Tucson, Arizona,
USA 85721

1 INTRODUCTION

During the last two years, a dramatic revision of our understanding of early stellar evolution has been occurring. Infrared,
optical, radio and in particular millimeter-wave observations have
revealed a new stage of early stellar evolution of unanticipated astrophysical importance. Energetic high velocity outflows of gaseous
material have been detected around numerous young stellar objects. For
example, large proper motions of optical Herbig-Haro objects have been
observed for two sources: L1551 and H-H 1 and 2 (Cudworth & Herbig 1981;
Herbig & Jones 1981). In both sources the Herbig-Haro objects are found
to be moving outward like bullets from a common center with velocities in
excess of 100 km s^{-1}! Similarly, centimeter-wave VLBI observations of
H_2O maser sources in Orion have detected proper motions of maser emitting
regions, indicating a large scale outflow of gas from an area near an
embedded infrared source known as IRC 2 (Genzel et al. 1981). Likewise,
near-infrared measurements of hydrogen recombination lines toward GL 490
and M17SW indicate linewidths in excess of 100 km s^{-1} suggesting that
ionized material close to the surface of these two stars is expanding
outward at high velocity (Simon et al. 1981). However, the most
intriguing, and perhaps most informative observations of energetic mass
outflow from around young stellar objects have been obtained at millimeter
wavelengths. Excessively broad emission line wings have been observed in
the fundamental rotational transition of CO toward at least twenty-six
infrared sources embedded in molecular clouds (Bally and Lada 1982).
The prototypical broad wing CO spectrum is that observed toward the
infrared cluster in Orion, a region coincident with the high velocity
mass outflow indicated by H_2O maser proper motions. Figure 1 shows a
spectrum of the submillimeter, J=3-2 line of ^{12}CO observed toward this
region with the 7-meter **aperture** Multiple Mirror Telescope in Arizona

Figure 1. Spectrum of the J=3-2 transition of ^{12}CO observed toward the
K-L region of Orion with the M.M.T. employing a novel beam combiner
designed by N. Erickson of the University of Massachusetts. High
velocity wings extend over a velocity range greater than 100 km s^{-1}
(adapted from Ulich et al. 1982).

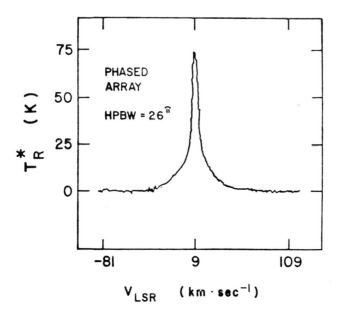

(Erickson et al., 1982; Ulich et al., 1982). Intense, broad ($\sim$100 km s^{-1})
emission wings are clearly evident. The angular extent of the high
velocity emission measured with the MMT is only about 40 arc seconds,
considerably less than the extent of the core or "spike" component of
the profile which extends over many degrees of sky (Kutner et al., 1977).
The observed high velocity molecular flow in Orion is highly supersonic
with velocities greatly exceeding the gravitational escape velocity in
the region. This fact coupled with the observations of H_2O maser proper
motions implies that a localized energetic outflow of molecular material
is occurring around the embedded infrared cluster in Orion.

The other twenty-five broad wing sources identified by Bally
and Lada (1982) all have line wing widths greater than 20 km s^{-1} at the
$T_r^* = 0.100$ K level when observed in the J=1-0 transition of CO with
the 7-meter Bell Laboratories millimeter-wave telescope. For all these
sources, the highest observed velocities greatly exceed those required
for gravitational confinement (Lada and Harvey, 1981). The size of a
typical high velocity molecular flow (hereafter HVMF) is about one parsec,

ten times larger than the extent of the Orion flow but still highly
localized when compared to the extents of the giant molecular clouds in
which most HVMFs are found.   Their short dynamical time scales (i.e.,
$\langle\tau\rangle = \langle R/V\rangle \sim 10^{4}$ yrs.) coupled with their high frequency of occurrence
suggest that HVMFs are the manifestation of a very common, but previously
unrecognized, stage of early stellar evolution (Lada and Harvey, 1981).

Millimeter-wave studies of the nature of the HVMFs have
revealed such unusual physical properties, that the HVMFs may be
considered among the most perplexing and perhaps important phenomenon
discovered in our galaxy.   In the remainder of this paper I will discuss
three of the most interesting characteristics of the HVMFs as presently
perceived from observations acquired by numerous astronomers during the
last two years.

2   SPATIAL STRUCTURE

Complete or nearly complete maps of CO emission have been
obtained for more than a dozen HVMFs.   Of these all but two (Orion A and
V645 Cyg) are reasonably well resolved.   Of the remaining sources one is
apparently spherically symmetric (S140, Lada and Wolf, 1982) while all
the others exhibit anisotropic angular distributions of the red and blue
shifted high velocity gas.   Figure 2 shows maps of seven HVMFs taken
from the survey of Bally and Lada (1982).   Remarkably most of these flows
appear bipolar in nature with red-shifted and blue-shifted emission
originating in **separate** spatial locations, more or less symmetrically
situated about an embedded infrared source.

The prototypical bipolar flow is L1551, the first bipolar
HVMF to be discovered (Snell, Loren and Plambeck 1980).   As mentioned
earlier, this flow contains Herbig-Haro objects with large proper motions.
These high velocity Herbig-Haro objects are coincident with the blue-
shifted lobe of high velocity CO gas and are moving outward away from a
common center very close to an embedded infrared source.   These observations clearly indicate that gas is being ejected from the vicinity of the
infrared source in two well collimated but oppositely directed streams or
jets.   Although the HVMFs in most sources are not as highly collimated as
the one in L1551, it is nevertheless quite surprising that bipolar
patterns persist at all, since the ambient medium into which these
streams are propagating is almost certainly inhomogeneous and of relatively high density (e.g. $10^{2} \leq n \leq 10^{5}$ cm^{-3}).   Collimating such high
velocity, hypersonic flows in such dense and clumpy environments presents

an important physical constraint on any mechanism which might be
forwarded to account for this phenomenon.

Figure 2. Maps of the distributions of integrated intensities in the red
and blue high velocity wings of ^{12}CO emission around seven sources.
Solid contours represent blue-shifted emission and dashed contours red-
shifted emission. Crosses locate the positions of embedded infrared
sources. (Solomon, Huguenin, and Scoville, 1981; Bally, 1982; Snell,
Loren, and Plambeck, 1980; Rodriguez, Ho, and Moran, 1980; Lada and
Harvey, 1981; Lada and Gautier, 1982).

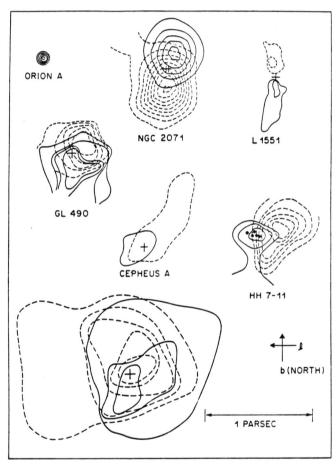

MAPPED HIGH VELOCITY MOLECULAR FLOWS

So far the only models proposed to explain the bipolar appearance of the flows have been qualitative ones. Snell, Loren, and Plambeck (1980) suggested that the jets in Ll551 could be the result of a circumstellar disk collimating an initially spherical stellar wind into well defined streams, with gas flowing outward along the poles of the disk. Observational evidence for the presence of such disks around the central sources of HVMFs has been discussed by Lada and Harvey (1981) and Bally (1982), but this evidence has been far from conclusive. Ho, Rodriguez and Moran (1982) have proposed a variant of this idea to explain the flow in Ceph A. In their model an initially spherical flow encounters large scale density clumps in the ambient cloud and becomes funneled into anisotropic streams. However, it is difficult to imagine how this model could produce highly collimated bipolar flows. Nonetheless, the model could be tested in other sources by comparing high angular resolution maps of density sensitive molecular transitions with spatial anisotropies in the high velocity CO gas. Clearly a more detailed observational and theoretical investigation of this problem is warranted.

3   ENERGETICS

The most problematic aspect of the HVMFs revealed by CO observations concerns their energetics. Since ^{13}CO emission is detect-able toward most flow sources (Lada and Harvey, 1981; Bally, 1982; Ho, Rodriguez and Moran, 1982; and Lada, 1982) relatively accurate masses can be obtained for the high velocity gas. Combined with measurements of the outflow velocities, these data can be used to estimate the kinetic energies and momenta of the high velocity gas. The kinetic energies of most flows are enormous (e.g., $10^{46} - 10^{47}$ ergs). If we compare the mechanical luminosities of the flows (i.e., $MV^3/2R$) with the total radiant luminosities of the central infrared sources, we find that the mechanical luminosities are very appreciable fractions (i.e., 0.01-0.1) of the radiant luminosities! This suggests a very efficient source of energy generation, almost as efficient as that responsible for producing the luminous energy from the embedded central infrared sources. A comparison of the radiant and mechanical luminosities also reveals a correlation between the two quantities as shown in Figure 3. Such a correlation suggests a common physical mechanism for producing the HVMFs and indicates that the energetics of the flows are somehow related to the luminosities or perhaps masses of the central objects.

Figure 3. The relation between the mechanical luminosity of the high
velocity molecular flows and the bolometric luminosities of the central
embedded infrared sources. Also plotted is the locus of points where the
flow mechanical luminosity equals the stellar radiant luminosity. (From
Bally and Lada 1982.)

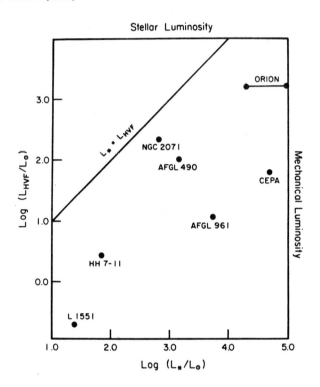

One possible mechanism for driving these outflows is radiation pressure (Phillips and Beckman, 1980). As shown in Figure 3, there
is certainly enough energy available in the radiation fields to drive the
HVMFs, although a very high conversion efficiency of radiant to mechanical
energy would be needed. The energetics of radiation driven winds would
also be expected to correlate with the luminosities of the central stars.
However, to properly assess the capability of such a mechanism, we need
to consider the momenta, more specifically, the forces required to drive
the HVMFs. The force required to drive a particular outflow is given
simply by the momentum of the flow divided by its dynamical age or
$F=\dot{M}V=MV/\tau=MV^2/R$, provided the flow has been steady over the dynamical
age. Figure 4 shows a comparison between such forces and the total

Figure 4. The observed relation between MV and central source bolometric luminosity for high velocity flows. (From Bally and Lada 1982.)

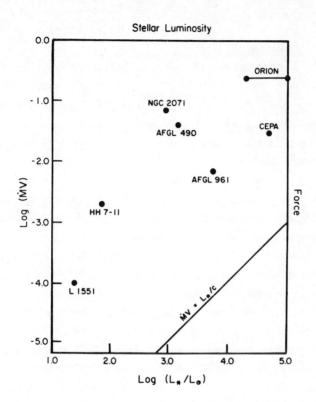

luminosities of the central sources. As is expected for radiation driven winds, the required driving force is correlated with luminosity. However, in all cases the force needed to drive the winds is orders of magnitude greater than that (L/c) which can be supplied by radiation pressure in an optically thin gas. This suggests that all the photons emitted by a central object would have to be scattered or absorbed 100 to 1000 times before escaping if radiation pressure is to be an important factor for driving the flows. Such a circumstance would appear unlikely.

Exactly what is driving the outflows is unclear. Perhaps semi-regular explosive outbursts, which are more energetic analogs of the FU Ori phenomenon are responsible. At the present time, however, the source of the energy for these molecular flows remains a mystery.

4   BIRTHRATE AND MOLECULAR CLOUD EVOLUTION

The astrophysical significance of this newly discovered phenomenon is only fully appreciated when one considers the high frequency of occurrence of HVMFs. Bally and Lada (1982) have estimated a HVMF birthrate of at least 3×10^{-4} $yr^{-1}kpc^{-2}$ in the local solar neighborhood, corresponding to a total galactic formation rate of 0.1 $yr.^{-1}$. This exceeds, by more than a factor of two, the formation rate of all stars with spectral types of A and earlier (Lada and Harvey 1981) and indicates the periods of energetic mass loss lasting approximately 10^4 years must be a common stage of early stellar evolution for stars of a wide range of spectral types. The identification of a new energetic stage of stellar evolution has important consequences not only for understanding of stellar evolution itself, but also for our understanding of the physics of molecular clouds within which stars are formed. The identification of a major source of internal energy for molecular clouds may lead to a solution of the long standing problem of the longevity of massive molecular clouds. It has long been known that thermal pressure in molecular clouds is too low to support them against gravitational collapse. The collapse time for molecular clouds is on the order of a few million years; yet molecular clouds appear to be at least ten times older than this. The longevity of giant molecular clouds can be explained if a sufficiently large internal source of energy exists that stabilizes the clouds against gravitational collapse. Norman and Silk (1980) recently proposed that winds from embedded T-Tauri stars supply the necessary internal energy to stabilize a molecular cloud. However, the kinetic energies of these winds were thought to be very low, and extremely high density of T-Tauri stars is necessary to support the clouds. The existence of the high velocity molecular flows which appear to occur around T-Tauri as well as B stars may remove the drawback of requiring an extremely high density of embedded stars. It is interesting to note that it would take about 2×10^6 years for all the high velocity sources to generate a mechanical energy equal to that observed for all the molecular gas in the Milky Way! This is comparable to the free-fall collapse time of the clouds. This suggests that high velocity molecular outflows play an important role in the support of molecular clouds. Alternatively, if molecular clouds are initially in hydrostatic equilibrium, then the HVFs would certainly contribute to their disruption (Lada and Gautier 1982).

5 CONCLUDING REMARKS

The discovery of bipolar, energetic mass outflow from around young stellar objects has opened up a new chapter in studies of early stellar evolution. A number of puzzling new problems have arisen with this discovery. As yet we do not understand the physics underlying this new phenomenon and although preliminary data are suggestive we are not entirely certain to what extent the energetic outflows effect molecular cloud evolution. Further studies of this phenomenon are clearly desired. Knowledge gained from such studies will not only enrich our understanding of stellar evolution, star formation and molecular clouds, but may also contribute to our general understanding of the physics of collimated mass flows and in doing so have an impact on other branches of astrophysics.

6 ACKNOWLEDGEMENTS

The author gratefully acknowledges the support of the Alfred P. Sloan Foundation and assistance provided by the staff of the Bell Telephone Laboratories in preparing this manuscript.

7 REFERENCES

Bally, J. 1982, Ap. J., (in press).
Bally, J. and Lada, C. J. 1982, Ap. J., (in press).
Cudworth, K. M. and Herbig, G. H. 1981, A.J., 84, 548.
Erickson, N., Goldsmith, P., Huguenin, R., Lada, C. J. and Ulich, B.
      1982, in preparation.
Genzel, R., Reid, M., Moran, J. M. and Downes, D. 1981, Ap. J., 244, 884.
Herbig, G. H. and Jones B. F. 1981, A.J., 86, 1232.
Ho, P. T. P., Rodriguez, L. F. and Moran, J. M. 1982, preprint.
Kutner, M. L., Tucker, K. D., Chin, G. and Thaddeus, P. 1977, Ap. J.,
      215, 521.
Lada, C. J. 1982, in preparation.
Lada, C. J. and Gautier, N. 1982, Ap. J., in press.
Lada, C. J. and Harvey, P. M. 1981, Ap. J., 245, 58.
Lada, C. J. and Wolf, G. 1982, in preparation.
Norman, G. and Silk, J. 1980, Ap. J., 238, 158.
Phillips, J. P. and Beckman, J. E. 1980, M.N.R.A.S., 193, 245.
Rodriguez, L. F., Ho, P. T. P. and Moran, J. M. 1980, Ap. J. (Letters),
      240, L149.
Simon, M., Righini-Cohen, G., Fischer, J. and Cessar, L. 1981, Ap. J.,
      251, 552.
Snell, R. L., Loren, R. B. and Plambeck, R. L. 1980, Ap. J. (Letters),
      239, L17.
Solomon, P. M., Huguenin, G. R., and Scoville, N. Z. 1981,
      Ap. J. (Letters), 245, L19.
Ulich, B., Lada, C. J., Erickson, N., Goldsmith, P., and Huguenin, R.,
      1982, In Advanced Technology Optical Telescopes: Proceedings
      of the S.P.I.E., ed. G. Burbidge, in press.

MOLECULAR CLOUDS IN THE GALACTIC NUCLEUS

R.J. Cohen
University of Manchester,
Nuffield Radio Astronomy Laboratories,
Jodrell Bank, Macclesfield, Cheshire
SK11 9DL, England

Abstract   First results are presented from a large-scale survey of OH in the galactic centre carried out at Jodrell Bank. The observations sample the region $-6\overset{\circ}{.}0 \leqslant \ell \leqslant 8\overset{\circ}{.}6$, $-2\overset{\circ}{.}0 \leqslant b \leqslant 1\overset{\circ}{.}6$. The concentration of massive molecular clouds in the galactic nucleus is twice as extensive as hitherto realized. New galactic centre clouds have been detected out to projected distances of 1 kpc either side of the centre. The outermost clouds lie out of the galactic plane in a tilted distribution, and reach z-distances of 200 pc above the plane at negative longitudes and below the plane at positive longitudes.

I will be describing observations made at a wavelength of 18 cm. My reason for presenting this work at a submillimetre conference is to encourage millimetre and submillimetre astronomers to observe the galactic centre. The centre is the only region of the Galaxy where the large-scale gas motions are not understood, and it is a region where millimetre and submillimetre observations have an important role to play in improving our understanding.

The unusual gas motions in the galactic nucleus were first investigated using the 21 cm line of neutral atomic hydrogen (HI) (Rougoor & Oort 1960). This shows strongly disturbed conditions in the central few kiloparsecs of the Galaxy. There are noncircular motions approaching $\pm$ 200 km s^{-1}, and the HI gas within 2.5 kpc of the centre occurs mainly out of the galactic plane in a tilted distribution (Cohen & Davies 1976; Burton & Liszt 1978; and references therein). Attempts have been made to model the gas motions in terms of rotation about the centre plus radial outflow, perhaps due to explosive events in the nucleus (e.g. van der Kruit 1971; Grape 1978), and in terms of elliptical stream-lines due to a central bar (e.g. Peters 1975). The two types of model appear to fit the observations equally well (e.g. Liszt & Burton 1980).

The 21 cm line has one great drawback for studying the galactic centre, namely the confusion produced by foreground gas near zero velocity.

Emission from this line-of-sight gas completely swamps any galactic centre emission at velocities less than $\pm$ 50 km s^{-1}, and thereby renders a large area of the galactic centre unobservable in HI. Molecular lines provide the opportunity to explore this previously hidden area. There is an enormous concentration of massive molecular clouds in the galactic nucleus. To an observer outside the Galaxy the nucleus would appear by far the most conspicuous molecular feature (cf. Cohen & Few 1981, their Fig. 7). Because of this there is almost no confusion between galactic centre features and foreground features in molecular spectra of the nucleus. The galactic centre features are readily distinguished by their strength and by their characteristically broad line profiles, which have velocity widths of typically 30 km s^{-1}. The molecular line studies carried out to date have shown an asymmetric concentration of molecular clouds distributed over longitudes $-1^{\circ} \leqslant \ell \leqslant 3^{\circ}$. The molecular layer is flatter than the HI layer, and taken as a whole it does not display the tilt seen in the HI, although some individual molecular features are slightly tilted (McGee et al. 1970; Cohen & Few 1981; and references therein).

I now want to present the first results from a new survey of OH in the galactic centre, carried out at Jodrell Bank this summer using the MK IA radio telescope. The survey considerably extends the sky coverage of previous molecular surveys of the centre. The instrumental details are summarized in Table 1. The OH mainlines at 1667 and 1665 MHz were observed simultaneously in the two hands of circular polarization. The lines appear primarily in absorption against the 18 cm continuum emission from the galactic nucleus and the galactic disk. This

TABLE 1	Instrumental Details
Lines observed	OH main lines at 1667 and 1665 MHz
Area surveyed	$-6\overset{\circ}{.}0 \leqslant \ell \leqslant 8\overset{\circ}{.}6$ $-1\overset{\circ}{.}0 \leqslant b \leqslant 1\overset{\circ}{.}0$ with extensions to $1\overset{\circ}{.}6$ and $-2\overset{\circ}{.}0$
MK IA Beamwidth	$0\overset{\circ}{.}17$ (10 arcmin)
Sampling interval	$0\overset{\circ}{.}2$ in ℓ and b
Velocity coverage	900 km s^{-1}
Velocity resolution	3 km s^{-1}
Detection level	0.2 K

circumstance would hinder the detection of molecular clouds located well
out of the galactic plane where the continuum background is weak. It is
important to search even further from the plane using the CO emission
lines, which do not suffer from this drawback. The 10 arcmin telescope
beam employed in the present work subtends a linear diameter of 30 pc at
the galactic centre, which is roughly the size of a typical molecular
cloud. Thus the present observations are mainly of value in studying the
large-scale aspects of the molecular cloud distribution.

Fig. 1 shows the area over which OH absorption was detected.
The upper diagram gives the peak absorption temperature, and is strongly
influenced by the distribution of the 18 cm continuum background. The
lower diagram gives the absorption as a percentage of the continuum, and
more fairly reflects the relative importances of the different molecular
clouds. The previously known concentration of molecular clouds shows up

Fig. 1   The upper diagram shows the strongest absorption
temperature measured in the 1667 MHz line, as a function of
position on the sky. The lower diagram shows the absorption
as a percentage of the 18 cm continuum. The dashed lines
indicate the boundary of the region surveyed. The telescope
beamsize is indicated at the lower right.

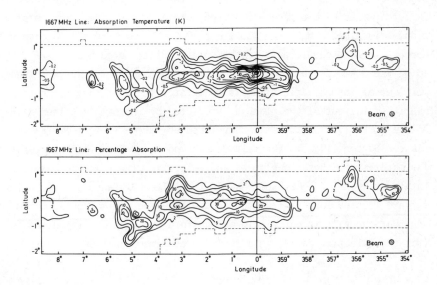

clearly between longitudes $\ell = 359°$ and $\ell = 3°$. In addition to this there is a more extensive outer distribution which is clearly tilted with respect to the galactic plane. Some of these outer clouds were also observed by Turner (1979) in his rather different OH survey. They extend to projected distances of ± 1 kpc from the galactic centre and ± 200 pc from the galactic plane. The projected tilt angle is similar to the value of $8°$ found by Cohen & Davies (1976) for high-velocity HI features. The clouds at negative longitudes give weaker OH absorption than the positive-longitude clouds. This might reflect a true deficiency in molecules: HI is also relatively deficient at these negative longitudes (cf. Cohen 1979, Fig.2). However at least some of the effect could be due to the position of the OH clouds relative to the 18 cm continuum emission. For example weaker OH absorption at negative longitudes was predicted by Cohen & Few (1976) on the basis of a symmetric bar model. Observations of CO emission might help to resolve the question.

Fig. 1 gives rather an oversimplified view of the nucleus, since it shows only the OH absorption of the strongest feature in a given direction. The full complexity of the region is apparent in Fig. 2, which is a preliminary longitude-velocity plot of all the main galactic centre features. Those at the highest velocities appear to be concentrations within already-known HI features, such as the "nuclear disk" described by Rougoor & Oort (1960), but there are numerous low-velocity features which have not been observed before in any line. The complex of clouds distributed near $\ell = 356°$, $b = +0°.8$, $V = -100$ km s^{-1} to 0 km s^{-1} appears to be a counterpart to the complex distributed near $\ell = 5°$, $b = -0°.6$, $V = 0$ km s^{-1} to 100 km s^{-1}. Although these complexes are located well out of the galactic plane it is interesting that they are not nearly so far out of the plane as predicted by the simple tilted-disk models of the gas distribution developed by Burton & Liszt (1978) and Liszt & Burton (1980). The tilted-disk models predict latitudes of $\pm 2°$ for material at these longitudes and velocities, whereas the observed latitudes are $\pm 0°.7$. It seems as if the tilted HI features modelled so far are located further from the centre than the new molecular features, and that the tilt is not simple and uniform but changes with distance from the galactic centre, in a fully three-dimensional way.

I have had time to touch on just a few of the many problems we face in understanding the galactic centre, which seems to get more complicated with each new set of observations that are made. I would like to finish by emphasizing the need for extensive millimetre and submillimetre observations. I hope that the observers in this room will remember Fig. 1 when they observe the galactic centre, and that they will not confine their attention to the two rather inconspicuous points Sgr A and Sgr B2.

Fig. 2 Schematic longitude-velocity plot of the galactic centre OH absorption features. Narrow-line features arising outside the nucleus are not shown. Small numbers indicate representative latitudes for the features. Features above the plane (b $\geq$ 0°) are shown black and those below the plane (b < 0°) are shown white. The thickness of a feature in this figure is proportional to the percentage absorption it produces in the 1667 MHz line.

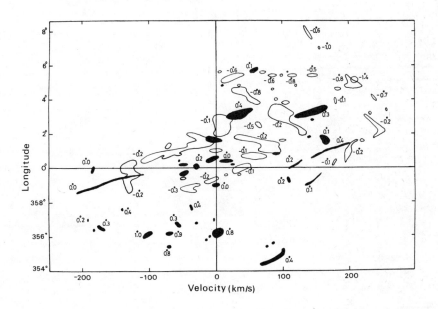

References

Burton, W.B. & Liszt, H.S. (1978). Astrophys.J., <u>225</u>, 815-842.
Cohen, R.J. (1979). IAU Symp.No. <u>84</u>, ed. W.B. Burton, pp.337-340.
         Dordrecht, Holland: Reidel.
Cohen,R.J. & Davies, R.D. (1976). Mon.Not.R.astr.Soc., <u>175</u>, 1-24.
Cohen, R.J. & Few, R.W. (1976). Mon.Not.R.astr.Soc., <u>176</u>, 495-523.
Cohen, R.J. & Few, R.W. (1981). Mon.Not.R.astr.Soc., <u>194</u>, 711-736.
Grape, K. (1978). Mon.Not.R.astr.Soc., <u>185</u>, 713-725.
Liszt, H.S. & Burton, W.B. (1980). Astrophys.J. <u>236</u>, 779-797.
McGee, R.X., Brooks, J.W., Sinclair, M.W. & Batchelor, R.A. (1970).
         Aust.J.Phys. <u>23</u>, 777-787.
Peters, W.L. (1975). Astrophys.J., <u>195</u>, 617-629.
Rougoor, G.W. & Oort, J.H. (1960). Proc.Nat.Acad.Sci., <u>46</u>, 1-13.
Turner, B.E. (1979). Astr.Astrophys.Suppl., <u>37</u>, 1-332.
van der Kruit, P.C. (1971). Astr.Astrophys., <u>13</u>, 405-425.

THE DISTRIBUTION OF CARBON MONOXIDE IN SPIRAL AND IRREGULAR GALAXIES

D.M. Elmegreen
Royal Greenwich Observatory, Herstmonceux Castle,
Hailsham, East Sussex, BN27 1RP, England

Increased sensitivity of radio receivers has made it possible to map the J=1-0 (2.6 mm) emission line of ^{12}CO in external galaxies. However, the typical beamsize used for extragalactic CO (65" with the NRAO 36' radio telescope) is sufficiently large that good linear resolution can be achieved only in the nearest galaxies. It is possible to improve the effective resolution by using overlapping beam positions and by utilizing the velocity information contained in the CO spectra. The 1 MHz filterbanks provide a velocity resolution of 2.6 km s^{-1}, which is comparable to the velocity dispersions which can be measured in neutral gas using 21 cm aperture synthesis (with typical beamsizes of 30"), or in ionized gas in H II regions using optical methods (with 1" resolution).

In principle, if the velocity field of a galaxy were completely mapped at high resolution, and the velocities of ionized, neutral, and molecular gas were the same, then the location of CO emission within a beam could be precisely determined. In practice, complete velocity fields are available in only a few cases; rotation curves have been made for many galaxies based on velocities measured only along the major axes. Since streaming or radial velocities are observed to be less than some 20 km s^{-1} in general, the extrapolation of a theoretical velocity field from the major axis rotation curve data is an adequate first approximation to the true velocity field. If the rotation curve includes data from both halves of the major axis, then problems with asymmetries in the major arms can be avoided. The distribution of the velocities in a galaxy are such that each observed CO feature

can be located in a region that is a thin, curved strip inside the whole area covered by a telescope beam. Figure 1
shows a schematic diagram of the process.

This technique has been applied to four SAB galaxies, NGC 157, NGC 2903, NGC 4321, NGC 5248; and to the Seyfert SA galaxy, NGC 1068 (Elmegreen & Elmegreen 1981). The
^{12}CO emission was observed in several positions in each galaxy, and the spectra were compared with the expected velocity
fields constructed from published rotation curves.

The CO emission from three of the above galaxies
will be considered in some detail. NGC 1068 is one of the
better studied galaxies at optical and CO wavelengths. Our
map of five overlapping beams formed a NE-SW by NW-SE cross
about the center of the galaxy, with a sixth overlapping beam
to the far SE. The emission is strongest in the center, where
a three-component line structure is seen. Based on the overlapping beam positions and the velocity data, a line at 1141
km s^{-1} originates close to the galaxian center, and the two
other peaks at 1040 km s^{-1} and 1233 km s^{-1} come from the respective vicinities of the NE and SW spiral arms, close to the
center. The intensity of the CO emission decreases with increasing radial distance. There is a strong dust lane in the

Figure 1 shows a schematic diagram of a two-armed
spiral galaxy with superposed radial velocity field
lines; typical line spacings are 20 km s^{-1}. The
circle represents a typical beamsize for CO emission.

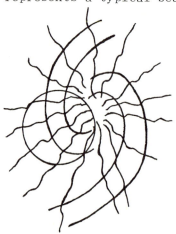

SE which appears to have some associated CO emission, while the patchy areas in the inner galaxy show no prominent CO.

CO emission was detected towards eight of the ten directions observed in NGC 4321, which provided essentially a complete single-beam map of the galaxy. The beams were centered along the two prominent spiral arms. The CO temperatures are close to 0.04 K in all cases except for the center, where it is 0.10 K. There is a slight tendency for the temperature to decrease with increasing distance along the arm. There also appears to be a correlation between the strength of the CO emission and the optical brightness of the clusters contained within the beam positions. The central velocities of the observed lines coincide with the expected velocities near the arm within each beam for the southern spiral arm. Near the northern arm, the CO velocities are as much as 50 km s^{-1} smaller than the expected velocities, which indicate negative peculiar velocities in those vicinities. In all cases in this galaxy, the CO emission appears to originate from the immediate vicinity of the bright spiral arms.

Eight overlapping beam positions essentially covered the bright optical arms and disk of NGC 5248. In two regions, including the center, CO emission is absent or weak (< 0.04 K). The strongest lines are from the two beam positions centered on the arms and adjacent to the central beam. The inferred association of CO emission with optical features is different in this galaxy compared with other galaxies in this study. Based on the beam positions alone, the CO emission is strongest towards the patchy, dusty interarm regions and weaker in the brighter arms. When the velocity data are examined, the same results are obtained. The line velocities corresponding to the expected velocities arise from the beam areas which are not directly on the spiral arms, but are the interarm regions near the arms. Large streaming motions can not account for the velocities in this case; they would be smaller velocities than the expected values, whereas the observed velocities are larger than expected for the arms.

It is evident that, while overlapping beams and detailed velocity maps are useful for pinpointing the location

of CO emission, a priori predictions of the CO locations are
not obvious. Whereas most of the emission in these galaxies
originated in the vicinity of dusty areas and star-forming
regions, the very regular galaxy NGC 5248 proved to be an exception. Such results are less surprising in view of a survey
of magellanic irregular galaxies (Elmegreen et al. 1980). In
that study, seven magellanic irregulars or magellanic spirals
were observed, and four were sampled in ten or more directions
for ^{12}CO emission (IC 1613, Ho II, IC 2574, NGC 6822). These
galaxies are sufficiently nearby that the beam corresponds to
the sizes of OB associations and clusters, H II regions, and
large dust clouds. Upper limits of 0.04 K for a 10 km s^{-1}
linewidth were obtained in all cases.

Several possible explanations might account for the
lack of CO detection; there might be lower CO temperatures
than the expected extrapolation from detections in spiral galaxies due to lower heating rates, or to lower CO abundances.
Lower CO temperatures might result from different cosmic ray
fluxes, which affect the temperatures of quiescent clouds.
Metallicities,observed to be low in irregular galaxies and rad-
ially variant in some spirals,could lead to lower CO abundance.
The abundance of CO could be less than expected even for normal C and O abundances if cosmic ray fluxes or dust-to-gas
ratios were low, since the former might slow ion-molecule reactions and the latter would increase the photodissociation of
CO by ultraviolet radiation. Finally, intense star-forming
activity could disrupt and dispel giant molecular clouds so
that CO would not be detected in the immediate vicinity of
young stars and dusty regions.

Thus, the distribution and abundance of CO in external galaxies is a complex and intriguing problem which can
only be fully understood and appreciated by making supplemen-
tary observations at other wavelengths, and by incorporating
theories of molecule, cloud, and star formation.

ACKNOWLEDGMENTS
The author was a visiting astronomer at the Royal

Greenwich Observatory during the preparation of this paper, and thanks the staff members for their hospitality.

REFERENCES

Elmegreen, B.G., Elmegreen, D.M., and Morris, M. (1980). On the abundance of carbon monoxide in galaxies: a comparison of spiral and magellanic irregular galaxies. Astrophysical J., 240, 455-463.

Elmegreen, D.M. & Elmegreen, B.G. (1981). CO observations of the SAB galaxies NGC 157, 2903, 4321, and 5248, and the Seyfert galaxy NGC 1068. Astronomical J., submitted.

OBSERVATIONS OF FAR-INFRARED EMISSION FROM LATE-TYPE GALAXIES

L. J Rickard
Department of Physics and Astronomy, Howard University,
Washington, DC, 20059, U.S.A.

P. M. Harvey
Department of Astronomy, University of Texas at Austin,
Austin, TX, 78712, U.S.A.

The galaxies with detected molecular components can be divided into two categories: those with strong central enhancements, and those without. Almost all of the galaxies with known molecular cloud components are of the first kind. The central sources have distinctive velocity widths, molecular-to-atomic mass surface density ratios, and so forth, suggesting that they represent a population of molecular clouds different from the GMCs that predominate in the disks. In order to get more information about these sources, we began a program of systematic observations of galactic centers, using a GaAs bolometer with effective wavelengths of 40, 50, 100, and 160 microns, operated from the NASA Gerard P. Kuiper Airborne Observatory.

We observed 25 spiral and Irr II galaxies, of which three are well-known sources, and 18 are new (unpublished) detections at 100 microns. The fluxes range from 3 Jy (M81, M101) to 97 Jy (NGC 2146); the majority of the new detections are below 20 Jy. The corresponding luminosities range from 3×10^7 L_0 (M81) to 5×10^{10} L_0 (NGC 2146). Fits to the spectra yield dust temperatures typically ranging from 31 K (NGC 5055) to 47 K (NGC 4736); NGC 4303 is unusual in that the dust temperature is only 16 K. The associated dust masses range from 3×10^3 M_0 (M81) to 3×10^7 M_0 (NGC 4303). Where CO data are available for the same angular region (50"), we find gas-to-dust ratios typically on the order of 300. We have combined these results with data for five other well-known IR-emitting late-type galaxies in order to do some statistical studies.

There is a suggestion that the 100-micron luminosity is correlated with the optical luminosity of the galaxy, although this may be mostly due to the presence of a few objects luminous in all bands (e.g., NGC 1068). There is a definite correlation of the 100-micron luminosity with the luminosity of the 2.6-mm CO line, where present. The trend persists if both quantities are scaled to the optical luminosity.

This may be understood as follows: The molecular clouds are the sites of formation for massive stars. Thus, more molecular gas should mean greater numbers of massive stars, and thus more heating of the coincident dust, and more radiation at 100 microns. The alternative, in which the presence of greater numbers of massive stars heat the dust, which then heats the gas (via collisions), and so results in a brighter (optically thick) CO line, does not seem to be correct. There is no correlation of the CO luminosity with the fitted dust temperature.

The 100-micron luminosity also correlates with the radio continuum luminosity, with the degree of concentration of radio emission to the galactic center, and with the integrated J-K colors. It does not correlate with total mass, with hydrogen mass, with the velocity width of the 21-cm hydrogen line, or with Hubble type.

A more detailed study of these data is currently being prepared for publication.

THE DISTRIBUTION OF CARBON MONOXIDE IN THREE FACE-ON GALAXIES

L. J Rickard
Department of Physics and Astronomy, Howard University,
Washington, DC, 20059, U.S.A.

P. Palmer
Department of Astronomy and Astrophysics, University of
Chicago, Chicago, IL, 60637, U.S.A.

Carbon monoxide emission at 2.6-mm wavelength provides the best currently available tool for tracing the distribution of molecular gas in galaxies. It is anticipated that studies of extragalactic CO will resolve a number of currently controversial questions about the nature and distribution of interstellar gas in galaxies. These questions (the form of the radial variation of the mass surface density of interstellar matter, the fraction of mass in molecular clouds, the degree of confinement of molecular gas to spiral arms) in turn relate to a variety of other issues, such as the efficiency of star formation and the lifetimes of molecular clouds. Furthermore, the special vantage of the extragalactic studies should provide valuable insights for the difficult study of interstellar gas in our own Galaxy.

We have recently made large-scale maps of the CO distributions of three nearly face-on late-type galaxies: M51, NGC 6946, and IC 342 (Rickard & Palmer 1981). These maps cover the bright optical disks of the galaxies, and thus provide the basic data for an immediate analysis of the overall properties of the molecular gas distributions.

The data were taken with the 36-foot telescope of the National Radio Astronomy Observatory in a sequence of observing runs starting in 1975 and ending in 1981. The observed positions were arranged in hexagonal grids, spaced by 1' (roughly equivalent to the full half-power beamwidth). Our initial analysis is based solely on maps of the integrated intensity under the line profile.

The CO emission appears to be confined to the bright optical disks. Enough points have been measured off the disks to verify that there is a strong drop-off at the optical "edges". By comparison, the extent of the HI distributions for these galaxies is much larger. If the data are averaged in radial bins, one finds essentially monotonic declines in the CO flux, dropping from the strong central enhancements to about

25% of the peak flux in a disk that seems to have fairly constant flux over an extent of about 10 kpc. The decline from the peak is the same for all three when viewed in angular distance, suggesting that the central sources are unresolved.

At a given radius, the variation of the actual fluxes about the azimuthal mean is comparable to the size of the mean itself. This makes it difficult to be precise about the form of the radial variation. We have explored some fits of model source distributions to the data, and find that statistical tests are unable to distinguish a model of nucleus plus flat disk from one in which the flux declines exponentially with radius.

Although there is considerable structure in the disks, there is little indication of any coherent nonaxisymmetric structures (i.e., spiral arms). The only arguable indications of the spiral patterns are the occurrences of peaks in the CO emission that can be matched up with specific HII conplexes. Admittedly, the 1' angular resolution is a limiting factor. But, it is clear that an arm/interarm contrast as large as that seen in the HII region distribution would show up in our data. We are currently doing statistical tests on comparisons of optical models for the spiral patterns with our data, in order to set limits on the possible contrast.

Smith (1982) has recently mapped M51 in the continuum at 170 microns. The far-infrared distribution is in general similar to the CO distribution, but it is difficult to match up specific features. It seems likely that a detailed intercomparison of these maps can yield some interesting information, such as the variation across the disk of the filling factor of molecular gas.

Clearly, the basic requirement for advancement in this area is better angular resolution. This could be achieved either with the large 3-mm telescopes now under construction, or with smaller instruments used for observations of the higher rotational transitions of CO in the submillimeter range.

Rickard, L.J & Palmer, P. (1981). Astron. Astrophys.,102, L13–L16.
Smith, J. (1982). Astrophys. J., in press.

SECTION III

Interstellar chemistry

THE INTERACTION BETWEEN CHEMISTRY AND ASTRONOMY

H.W. Kroto
School of Chemistry and Molecular Sciences, University of
Sussex, Brighton, BN1 9QJ, England

Abstract. It is only in the last decade that the true extent
of the molecular components of the interstellar medium (ISM)
has become evident. Suddenly Chemistry has taken on a new
significance for Astronomy in so far as it affects the study
of the ISM at low and moderate temperatures and in particular
those regions where stars are in the process of being born.
This article deals with this recent synergistic combination
of Chemistry and Astronomy. Herzberg (1950) has discussed
the particular interaction between Spectroscopy and Astronomy.
Here we consider recent results from the perspective of
molecular science in general.

OPTICAL OBSERVATIONS

There were a number of important conclusions to be drawn from
the first spectroscopic observations in 1936-1940 of interstellar CH, CH^+
and CN. The mere fact that molecules existed in such low pressure
regions raised questions about how they were formed in the first place,
apparently under conditions in which only binary collisions could occur.
Indeed these observations instigated theoretical studies of interstellar
molecule formation and the abundances of CH and CH^+ still present
problems (Dalgarno 1976). It is thought that radiative association in
which two molecules collide and the excess energy is carried away by a
photon is one of the initial steps. The molecules can be used to probe
the physical and chemical conditions in the ISM. A nice example of the
sort of information available at that time is highlighted by the
interstellar spectrum of CN. The laboratory spectrum (Pearse and Gaydon
1963, Herzberg 1950) consists of many transitions because under the
excitation conditions a large number of states are populated. In
contrast the interstellar spectrum (Adams 1941, and Thaddeus 1972) shows
only three lines because only the two lowest rovibronic levels are
significantly populated.

One can determine the interstellar rotational excitation temperature for CN very neatly from the ratio of the heights of the lines. Thaddeus (1972) gives a value of 2.99 ± 0.10 K.   Interestingly a temperature of 2.3 K had originally (in 1941) been determined from the intensities of these lines and had caused Herzberg (1950) to say it "... has of course only very restricted meaning." The restricted meaning is now recognised as one of the most important measurements of the three degree background which is perhaps the most persuasive piece of evidence in favour of the Big Bang theory of the origin of the Universe.

The early optical experiments were, of course, restricted to ground based observations and recent experiments have been carried out above the atmosphere allowing ultraviolet and vacuum ultraviolet experiments to be carried out. Using a rocket-borne spectrometer absorption of starlight by diffuse interstellar clouds has finally revealed the most important interstellar molecule of all, H_2.   The Lyman Bands of H_2 near 1092 $\overset{o}{A}$ have been detected by Carruthers (1971) and the Copernicus Satellite brought back further detailed results.

Another important molecule has been added to the list of interstellar molecules by optical measurements. The molecule C_2 was detected by ground based spectroscopy at $\sim$ 8750 $\overset{o}{A}$, just in the infrared, by Souza and Lutz (1977).

The diffuse interstellar lines which lie between 4400 and 6800 $\overset{o}{A}$ were also detected during the same period that CH, CH^+ and CN were first observed. These are a set of broad diffuse absorption features, the first of which were detected by Merrill (1934). Since then more have been detected and there are now some 39 features whose characteristics have been reviewed by Wu (1972), Herbig (1975), Smith, Snow and York (1977). They have now puzzled astronomers and spectroscopists for nearly five decades. It has been suggested that the lines belong to a molecule which might have predissociated structure or unresolved rotational structure (Herzberg 1965, 1967).

It has also been suggested that the carrier is (or are) a constituent of solid grains perhaps matrix trapped impurities (Merrill and Wilson 1938). A more recent suggestion by Douglas (1977) is that the bands belong to various C_n chain molecules. Suffice it to say that the

identity is really no more certain today than it was when they were first
detected and this puzzle remains one of the outstanding unsolved
mysteries.

RADIO OBSERVATIONS

The detection of the 21 cm line of hydrogen in 1951
revolutionised galactic astronomy and for the first time the overall
structure of our galaxy became apparent. Maps of H atom densities were
made which delineated arm-like features similar to those in other spiral
galaxies. An interesting point is that in studies of dark clouds such
as Bok globules the H atom intensity tended to increase towards the
feature and then decrease towards their interiors (Dickman 1977) a clear
sign that in these clouds the hydrogen was in molecular form.

The existence of free radicals implied that species such as
OH and NH might also be detectable and in fact OH transitions were
detected by Weinreb et al. (1963). The observed intensities are
anomalous indicating non-equilibrium excitation. Often very high
brightness temperatures are observed from very compact regions indicating
that special excitation conditions are pumping maser emission (Moran
1976). A most important early result was the occasional observation of
an anticorrelation with H 21 cm radiation and a correlation with dust
clouds, the first real evidence that molecules were indeed lurking in
these secluded regions of the ISM.

The ammonia molecule was the decisive key (Cheung et al.
1968, 1969) which opened up this subject in 1968. It is an oblate
symmetric top whose levels are split by inversion giving rise to many
transitions which lie near 24 GHz.

A valuable aspect of the inversion spectrum is the proximity
of the frequencies with different excitation energies. This allows
the relative intensities to be studied with one telescope and the
excitating conditions to be readily derived. Thus NH_3 is quite a useful
molecule to use to study collision and radiative energy distribution
mechanisms. Some care however is necessary as the results on NH_3 in
SgrB2 of Winnewisser, Churchwell and Walmsley (1979a) indicate. Their
data show that complicated velocity structure and spatially inhomogeneous
excitation conditions exist in this important molecular cloud.

As well as detecting NH_3 Cheung et al. (1969) also detected H_2O. This particular line is often detected in association with the OH masers and also often shows maser action itself.

Recently a low excitation energy line of H_2O has been detected at 183 GHz using a 91 cm radio telescope on the Kuiper flying lab., a C141 Lockheed transport which can fly at sufficiently high altitude to reduce atmospheric absorption for this (Waters et al. 1980).

Formaldehyde, H_2CO, which was discovered in 1969 by Snyder et al. (1969) was the first of a series of small organic molecules to be detected. The transitions are usually detected in absorption and in fact are often observed in absorption against the 3K background. This originally perplexing observation indicates that there is some mechanism possibly a collisional process (Townes and Cheung 1969, Garrison et al. 1975) cooling the molecules or more correctly the associated levels below the ambient temperature.

The early period around 1970 saw the development of sensitive receivers and detectors able to observe at higher and higher frequencies. This improvement enabled the Bell Laboratories group to detect $J=1\rightarrow0$ line at CO at 115 GHz (Wilson, Jefferts and Penzias 1970). CO is now known to be the most abundant molecule after H_2, some 1000 times more abundant than any other species so far detected. A great deal of work has been carried out with this line and the most important result is that the ratio n_{CO}/n_{H_2} $\sim$ 10^{-4} and appears to be very constant throughout the galaxy.

Thus the best way of determining the mass of interstellar H_2 in the galaxy, its location and its density distribution is via the ubiquitous CO emission (Gordon and Burton 1979).

Many stable organic and inorganic species have been observed. Some, which are really intermediate between inorganic and organic such as NH_2CN, HCOOH and HNCO have also been detected. In fact it is highly likely that most molecules involving C, N, O and H which have any reasonable stability exist in the ISM and in Table I a list of known species is presented.

UNSTABLE SPECIES

The conditions in the ISM ensure that once formed, molecules which react rapidly in the laboratory, may be quite stable. After the initial phase of molecule detection which was in general governed by the availability of microwave data on known (mainly everyday) species an interesting interaction between astronomy and molecular spectroscopy began to appear. The realisation that interstellar chemistry favoured some unusual and interesting species instigated and certainly encouraged the development of new spectroscopic techniques to study new molecules as well as others which are very difficult to handle in the laboratory.

A possible identification of interstellar glycine, H_2NCH_2COOH, the simplest amino acid, has been reported by Hollis et al. (1980). Microwave measurements on glycine originally yielded the spectrum of the conformer I in which the hydroxyl H atom was H bonded via the N lone

I   II

pair as shown (Brown et al. 1978; Suenram and Lovas 1978). Subsequent radio searches for this species proved negative. Theoretical studies of the glycine potential surface by Vishveshwara and Pople (1977) and Sellers and Schäfer (1978) however indicated that a second conformer (II) should be more stable than I by ~ 350 cm^{-1}. Suenram and Lovas (1980) subsequently detected the microwave spectrum of conformer II and on the basis of their laboratory measurements have identified one possible interstellar line (Hollis et al. 1980).

An intriguing series of long linear molecules the cyano-polyynes has been discovered. Hydrogen cyanide, $HC \equiv N$, and cyanoethyne (cyanoacetylene), $HC \equiv C-C \equiv N$, which can be considered to be the first and second members in the series were discovered by Snyder and Buhl (1971) and Turner (1971) respectively. The next member of the series, $HC \equiv C-C \equiv C-C \equiv N$, was synthesised and studied by Alexander, Kroto and Walton (1976) and the laboratory frequency then used to detect the molecule in SgrB2 (Avery et al. 1976). This discovery stimulated the laboratory synthesis of the next member of the series, $HC \equiv C-C \equiv C-C \equiv C-C \equiv N$, (Kirby, Kroto and Walton 1980) and its subsequent detection in the cold quiet cloud in Taurus TMC1 by Kroto et al. (1978).

Having detected HC_7N the quest for the next member of the series, HC_9N, was initiated both experimentally and theoretically. A neat and simple extrapolation technique discovered by Oka (1978) was used to predict the B_0 value of HC_9N. Transitions were in fact detected, remarkably close to the predicted positions, by Broten et al. (1978) and in fact HC_9N was found in surprisingly high abundance in the cloud TMC1. HC_5N and HC_7N have also been detected in IRC+10216 by Winnewisser and Walmsley (1978).

This intriguing family of molecules has simple transitions which march systematically across the radio spectrum depositing convenient almost equidistantly spaced lines within the range of most receivers. The transitions of a linear molecule like HC_5N stand out like beacons amongst the multitude of lines associated with rotations of other molecules. This has resulted in the active study of the cyanopolyynes with a view to obtaining information on the structure of dense clouds, number densities, isotope fractionation and physical conditions (Avery 1980, Tölle et al. 1981 and Benson and Myers 1980).

The detections of thioformaldehyde, $CH_2=S$, and methanimine, $CH_2=NH$, highlight some interesting aspects of the interaction between chemistry and astronomy in developing new techniques and results. These two isovalent species tend to be rather unstable under the usual laboratory conditions though they can be studied readily under the low pressure conditions which are necessary for microwave measurements. They were first produced by high temperature chemistry in flow systems by Johnson, Powell and Kirchhoff (1971) and Johnson and Lovas (1972) respectively. On the basis of the frequencies determined in the laboratory for these new molecules the interstellar signals were detected (Sinclair et al. 1973, Godfrey et al. 1973).

A few free radicals have been searched for and detected in the ISM by Radioastronomy. Although no laboratory radio spectra were known at the time Jefferts, Penzias and Wilson (1970) were able to detect interstellar CN on the basis of the approximate rotational and spin-doubling constants derived from the optical spectrum (Poletto and Rigutti 1965). The laboratory microwave spectrum has now been observed (Dixon and Woods 1977).

After a considerable amount of laboratory work culminating in an interferometric determination of the Λ-doubling frequency by Baird and Bredohl (1971) the radio spectrum of CH was finally detected by Rydbeck, Ellder and Irvine (1974) and Turner and Zuckerman (1973).

The microwave spectrum of the formyl radical HCO, detected in the laboratory by Bowater, Brown and Carrington (1971) and Saito (1972), has also been observed in the ISM by Snyder, Hollis and Ulich (1976).

The first ion studied by microwave spectroscopy was CO^+ which was detected by Dixon and Woods (1975) in a discharge experiment. The resulting frequency measurements have been used to search for CO^+ in the ISM and a possible assignment has recently been made for this elusive species by Erickson et al. (1981).

As the searches for interstellar lines continued, occasionally lines were detected accidentally which were not assignable to known species. The strengths of some of these U-lines (Snyder 1972) were so great that they could only belong to small and stable (at least in the ISM) species. Klemperer (1970) suggested one of the lines detected by Buhl and Snyder (1970) might belong to HCO^+. This intriguing conjecture has subsequently been verified by laboratory experiments carried out by Woods et al. (1975).

A search for interstellar HNC by Buhl and Snyder (1971, 1972) turned up a strong line at 90.7 GHz as a possible candidate. Theoretical calculations (Booth and Murrell 1972, Barsuhn 1972) and the detection of the laboratory spectrum simultaneously by Saykally et al. (1976), Blackman et al. (1976) and Cresswell et al. (1976) have nicely confirmed this assignment.

A group of three closely spaced U-lines first detected by Turner (1974) was finally assigned to the $J=1\rightarrow0$ transition of the linear protonated nitrogen ion HN_2^+ by Green, Montgomery and Thaddeus (1974).

A fourth molecule, the free radical C_2H, was also detected accidentally and assigned by similarly neat detective work. In this case a group of four lines was recognised by Tucker, Kutner and Thaddeus (1974) as hyperfine structure due to spin-rotation and nuclear spin/ electron coupling. The parameters which scale the latter interaction had already been determined by Graham, Dismuke and Weltner (1974) in matrix isolation e.s.r. experiments and convincingly verified the assignment.

Three of these interesting molecules: HCO^+, HNC, as well as HN_2^+ (Saykally 1976a) (Woods et al. 1975) have been observed by Woods and co-workers in the laboratory. The detection of these four unusual species is one of the more elegant stories in this always interesting field. These four molecules perhaps more than any others, have lent great weight to the claims of the ion-molecule aficionados that their mechanism is the major one in the ISM.

Somewhat later Guélin and Thaddeus (1977) and Guélin, Green and Thaddeus (1978) were able to add two more exciting species to this list, CCN and CCCCH. These are both free radicals produced by abstracting H atoms from cyanoethyne and butadiyne respectively.

Very recently Thaddeus, Guélin and Linke (1981) have also presented a strong evidence for the assignment of three U lines as the J=3, 4 and 5 transitions of a linear molecule which is either $HOCO^+$ or HOCN.

Winnewisser, Churchwell and Walmsley (1979) have surveyed the work on the various U-line and free radical detections involving HCO^+, HN_2^+, HNC, HC_2, HC_4, C_3N, CN, CH and NS. Wilson (1980) has reviewed the applications of theoretical calculations to these and future assignments.

The list of unassigned lines is growing with time and attempts have been made to assign some of them (see for instance Rodriguez Kuiper et al. 1977; Lovas 1974 and Kroto et al. 1978). A listing has been compiled by Turner (1979) of 120 or more.

INTERSTELLAR CHEMISTRY

Ever since the detection of CH, CH^+ and CN the problem of how such molecules could be formed and survive in the ISM has been a field for study. The detections since 1968 (Table I) have injected the study of interstellar chemistry with a new lease of life. A general review has been presented by Watson (1976).

The three main processes are: reactions on grain surfaces, 2-body gas phase ion molecule reactions, circumstellar shell formation followed by ejection into the ISM. Of course all three may be important and the balance is certainly not at all clear.

TABLE I   OBSERVED INTERSTELLAR MOLECULES[†]

H_2	H_2O	NH_3	CH_4	CH_3NH_2	CH_3OCH_3
OH	H_2S	H_2CO	CH_2NH	CH_3CHO	CH_3CH_2OH
CH	HCO	H_2CS	NH_2CN	CH_3C_2H	HC_7N
CH^+	HCO^+	H_2C_2	$HCOOH$	CH_2CHCN	
C_2	HCN	$HNCO$	HC_3N	HC_5N	
CN	HNC	$HNCS$	HC_4		HC_9N
CO	HNO	C_3N		$HCOOCH_3$	
CO^+	HN_2^+		CH_3OH		
CS	HC_2		CH_3SH	CH_3C_3N	
NS	SO_2		NH_2CHO		
SO	OCS				
SiO					
SiS					

[†]also $HOCO^+$ or $HOCN$ (see text). Molecules observed in circumstellar shells are also included.

TABLE II   FRACTIONAL ABUNDANCES OF INTERSTELLAR MOLECULES[†]

R	Species				
1	H_2				
10^{-4}	CO				
10^{-6}	HCN	HNC			
10^{-7}	OH	CS	SO	SiO	SiS
	HCO^+	N_2H^+	C_2H	SO_2	CH_3OH
10^{-8}	CH	CN	NS	HCO	
	H_2S	OCS	H_2CO	HC_3N	
10^{-9}	$HNCO$	CH_3C_2H			
10^{-10}	H_2CS	CH_2NH	$HCOOH$	NH_2CHO	
	CH_3CHO	CH_3NH_2	CH_2CHCN	HC_5N	
	CH_3C_3N	$HCOOCH_3$	CH_3CH_2OH	CH_3OCH_3	

[†]This table lists $R = n_x/n_{H_2}$ for the denser clouds where $n_{H_2} \sim 10^3\text{-}10^6$ cm^{-3}. Taken from Huntress (1977b).

In general two colliding atoms (i.e. two H atoms) cannot stick together unless the excess kinetic energy is taken away by a third body. This may be a simultaneously colliding third body, an emitted photon or an electron. Thus the big problem is to find a way for H_2 molecules to form in the first place. They cannot form by radiative association as only quadrupole emission, which is very weak, is allowed. The most favoured mechanism for forming H_2 is the grain surface catalysed reaction (McCrea and McNally, 1960). The hydrogen atom attaches to a grain and migrates over the surface until it reacts with a second H atom forming H_2 which then detaches.   CO and N_2 may be able to evaporate but many of the recently detected species such as CH_3CH_2OH or even H_2O might have very great difficulty at the low temperatures in the ISM.

As far as grain chemistry in general is concerned the unknown surface composition is a major problem. Indeed it is fair to say that little if anything is known about their chemical composition. Laboratory investigations which should shed light on this field are being pursued (Hagen, Allamandola and Greenberg, 1979, Allamandola et al.,1979).It is possible that some chemical mechanisms such as exothermic bond formation may heat up small grains sufficiently for molecules to evaporate or cause localised heating of larger grains. It is also possible that photons and X or cosmic rays or electrons may heat the area where the molecule is stuck sufficiently to cause it to unstick.

A large number of possible grain surface reactions have been discussed by Allen and Robinson (1977) as possible routes to interstellar species.

Exothermic gas-phase ion-molecule reactions, many of which have zero activation energies and large reaction cross-sections, offer particularly attractive routes to interstellar molecules. There are several major reviews highlighting various aspects of the chemistry in detail. Solomon and Klemperer (1972), Dalgarno and Black (1976) and Dalgarno (1976) have discussed the chemistry of diffuse clouds, Herbst and Klemperer (1973) that of dense clouds. There are also useful reviews by Huntress (1977a), McDaniel et al. (1976), Watson (1977, 1978) and by Smith and Adams (1980a).

The ionisation of species such as H, H_2 and He by photons in diffuse clouds and cosmic rays in dense clouds is passed on to less abundant atoms such as O etc. by chains of 2-body transfer reactions which in turn lead to some of the small polyatomic molecules detected in the ISM. Such processes can proceed at every collision.

For instance H_2O may be formed via a sequence such as

$$He \xrightarrow{c.r.} He^+ \xrightarrow[+e]{O} O^+ \xrightarrow[+He]{H_2} OH^+ \xrightarrow[+H]{H_2} OH_2^+ \xrightarrow[+H_2]{H_2} OH_3^+ \xrightarrow[+H]{e^-} OH_2{}_{+H}$$

which involves cosmic ray ionisation, electron transfer, three exchange reactions and a final dissociative e^- recombination.

In the diffuse clouds photons as well as cosmic rays can penetrate and inject the energy which is converted to chemical energy. However, these are relatively unshielded regions and so some form of kinetic balance is set up between the formation and photodissociation processes which allows only the smaller (mainly diatomic) and less readily dissociated species such as CO to survive and build up in significant concentrations.

In denser clouds only high energy cosmic rays can penetrate and inject the requisite energy and once formed the molecules are so well protected by their grain colleagues that they will last indefinitely.

Apart from the fact that many of the small molecules can be accounted for by these processes the observation of HCO^+, HN_2^+ and HNC are rather convincing indicators that these reactions are significant. In addition the departure of the apparent interstellar isotope ratios from the cosmic ratios seems to be nicely explained by ion-molecule isotope fractionation.

The laboratory work of Smith and Adams (1980b) shows how fractionation can neatly explain the observed interstellar $^{12}CO/^{13}CO$ ratio, so providing strong arguments in favour of ion-molecule reactions.

Smith and Adams (1977) have determined many of the rate constants required for a detailed analysis of ion-molecule chemistry using the SIFT (selected ion flow tube) technique. Huntress (1977a) has also discussed complementary data using the ICR (ion cyclotron resonance) technique. The effective temperatures of the reactants in the ICR method may be quite high and thus the data may not apply directly to the low temperature ISM.

Prasad and Huntress (1980) have recently presented the results of a very large computer study of the chemistry of the ISM. It is still not at all obvious how molecules such as HC_9N can be produced in cold clouds by a sequence of two-body coupling reactions. Schiff et al. (1980) have suggested a possible sequence of steps. It may be that radiative association and neutral-neutral processes may become significantly more important for larger molecules.

The third process which must be taken very seriously is that of molecule formation in high temperature, high density, circumstellar envelopes followed by their subsequent ejection into the general ISM.

CONCLUSIONS

The impact of Chemistry on Astronomy has in the last decade become excitingly apparent. The cross-fertilisation has enabled us to penetrate dark clouds and chart unknown regions where the very low energy processes which are the first phases of star formation are taking place. It still remains to be seen how much of the ISM, or indeed the Universe, is in molecular form.

As far as chemistry is concerned many new experimental techniques have been and are being developed due to the charismatic appeal of astrophysically significant problems.

Adams, W.S. (1941). Ap. J., 93, 11.
Alexander, A., Kroto, H.W., and Walton, D.R.M. (1976). J. Mol. Spectrosc.,
         62, 175.
Allamandola, L.J., Greenberg, J.M. and Norman, C. (1979). Astron. and
         Astrophys., 77, 66.
Allen, M. and Robinson, G.W. (1977).   Ap. J., 212, 396.
Avery, L.W. (1980).   Interstellar Molecules, ed. B.H. Andrews , p.47. IAU
Avery, L.W., Broten, J.M., MacLeod, J.M., Oka, T. and Kroto, H.W. (1976).
         Ap. J. 205, L173.
Baird, K.M. and Bredohl, H. (1971).   Ap. J. 189, L83.
Barsuhn, J. (1972). Ap. J. 12, L196.
Benson, P.J. and Myers, P.C. (1980).   Ap. J. 242, L87.
Blackman, G.L., Brown, R.D., Godfrey, P.D. and Gunn, H.I. (1976).   Nature,
         261, 395.
Booth, D. and Murrell, J.N. (1972).   Mol. Phys., 24, 1117.
Bowater, I.C., Brown, J.M. and Carrington, A. (1971).   J. Chem. Phys.,
         54, 4957.
Broten, N.W., Oka, T., Avery, L.W., MacLeod, J.M. and Kroto, H.W. (1978).
         Ap. J. 223, L105.
Brown, R.D., Godfrey, P.D., Storey, J.W.V. and Bassez, M-P. (1978).
         J.C.S. Chem. Comm., 547.
Buhl, D. and Snyder, L.E. (1970). Nature, 228, 267.
Buhl, D. and Snyder, L.E. (1971).   Bull. Am. Astron. Soc., 3, 388.
Buhl, D. and Snyder, L.E. (1972).   Ann. N.Y. Acad. Sci., 194, 17.
Carruthers, G.R. (1971).   Ap. J. 166, 348.
Cheung, A.C., Rank, D.M., Townes, C.H., Thornton, D.C., and Welch, W.J.
         (1968), Phys. Rev. Lett., 21, 1701.
Cheung, A.C., Rank, D.M., Townes, C.H., Thornton, D.D. and Welch, W.J.
         (1969). Nature, 221, 621.
Cresswell, R.A., Pearson, E.F., Winnewisser, M. and Winnewisser, G. (1976).
         Z. Naturforsch, 31a, 221.
Dalgarno, A. (1976).   Frontiers of Astrophysics, ed. E.H. Avrett, p.352.
         Harvard Univ. Press, Cambridge, Mass.
Dalgarno, A. and Black, J.H. (1976). Reports on Prog. in Physics, 39, 573.
Dickman, R.L. (1977). Sci. Am. 236(6), 66.
Dixon, T.A. and Woods, R.C. (1975).   Phys. Rev. Lett., 34, 61.
Dixon, T.A. and Woods, R.C. (1977).   J. Chem. Phys., 67, 3956.
Douglas, A.E. (1977).   Nature, 269, 130.
Erickson, N.R., Snell, R.L., Loren, R.B., Mundy, L. and Plambeck, R.L.
         (1981).   Ap. J. 245, L83.
Garrison, B.J., Lester, W.A., Milles, W.H., and Green, S. (1975).
         Astrophys. J., 200, L175.
Godfrey, P.D., Brown, R.D., Robinson, B.J. and Sinclair, M.W. (1973).
         Astrophys. Lett., 13, 119.
Gordon, M.A. and Burton, W.B. (1979).   Sci. Am. 240(5), 54.
Graham, W.R.M., Dismuke, K.I. and Weltner, W. (1974).   J. Chem. Phys.,
         60, 3817.
Green, S., Montgomery, J.A. and Thaddeus, P. (1974).   Ap. J. 193, L89.
Guélin, M., Green, S. and Thaddeus, P. (1978).   Ap. J., 224, L27.
Guélin, M. and Thaddeus, P. (1977).   Ap. J., 212, L81
Hagen, W., Allamandola, L.J. and Greenberg, J.M. (1979).   Thermodynamics
         and Kinetics of Dust Formation in the Space Medium, ed.
         P. De and G. Arrhenius.
Herbig, G.H. (1975).   Ap. J., 196, 129.
Herbst, E. and Klemperer, W. (1973).   Ap. J., 185, 505.
Herzberg, G. (1950).   Molecular Spectra and Molecular Structure Vol. 1;
         Spectra of Diatomic Molecules, 2nd ed. Van Nostrand.

Herzberg, G. (1965). J. Opt. Soc. Am., 55, 229.
Herzberg, G. (1967). IAU Symposium, 31, 91.
Hollis, J.M., Snyder, L.E., Suenram, R.D. and Lovas, F.J. (1980). Ap. J. 241, 1001.
Huntress, W.T. (1977). Ap. J. Suppl. 33, 495.
Huntress, W.T. (1977). Chem. Rev., 295.
Jefferts, K.B., Penzias, A.A. and Wilson, R.W. (1970). Ap. J., 161, L87.
Johnson, D.R. and Lovas F.J. (1972). Chem. Phys. Lett., 15, 65.
Johnson, D.R., Powell, F.X. and Kirchhoff, W.H. (1971). J. Mol. Spectrosc., 39, 136.
Kirby, C., Kroto, H.W. and Walton, D.R.M. (1980). J. Mol. Spectrosc., 83, 261.
Klemperer, W. (1970). Nature, 227, 1230.
Kroto, H.W., Kirby, C., Walton, D.R.M., Avery, L.W., Broten, N.W., MacLeod, J.M. and Oka, T. (1978). Ap. J., 219, L133.
Kroto, H.W., Murrell, J.N., Al-Derzi, A. and Guest, M.F. (1978). Ap. J., 219, 886.
Lovas, F. (1974). Ap. J., 193, 265.
McCrea, W.H. and McNally, D. (1960). Mon. Not. Roy. Astron. Soc., 121, 238.
McDaniel, E.W., Cermák, K.V., Dalgarno, A., Ferguson, E.E. and Friedman, L. (1976). Ion-molecule Reactions, Wiley, Interscience, N.Y.
Merrill, P.W. (1934). Pub. A.S.P., 46, 206.
Merrill, P.W. and Wilson, O.C. (1938). Ap. J. 87, 9.
Moran, J.M. (1976). Frontiers in Astrophysics, ed. E.H. Avrett, p. 385, Harvard Univ. Press, Cambridge, Mass.
Oka, T. (1978). J. Mol. Spectrosc., 72, 172.
Pearse, R.W.B. and Gaydon, A.G. (1963). The Identification of Molecular Spectra, Chapman and Hall.
Poletto, G. and Rigutti, M. (1965). Nuovo Cimento, 39, 515.
Prasad, S.S. and Huntress, W.T. (1980). Ap. J., 239, 151; Ap. J. Suppl., 43, 1.
Rodriguez Kuiper E.N., Kuiper, T.B.H., Zuckerman, B. and Kakar, R.K. (1977). Ap. J., 214, 394.
Rydbeck, O.E.H., Ellder, J. and Irvine, W.M. (1974). Astron. and Astrophys., 33, 315.
Saito, S. (1972). Ap. J., 178, L95.
Saykally, R.J., Dixon, T.A., Anderson, T.G., Szanto, P.G. and Woods, R.C. (1976). Ap. J., 205, 2101.
Saykally, R.J., Szanto, P.G., Anderson, T.G. and Woods, R.C. (1976a). Ap. J., 204, L143.
Schiff, H.I., McKay, G.I., Vlachos, G.A. and Bohme, D.K. (1980). IAU Symposium, 87, 307.
Sellers, H.L. and Schäfer, L. (1978). J.A.S., 100, 7728.
Sinclair, M.W., Fourikis, N., Ribes, J.C., Robinson, B.J., Brown, R.D. and Godfrey, P.D. (1973). Aust. J. Phys., 26, 85.
Smith, D. and Adams, N.G. (1977). Ap. J., 217, 741.
Smith, D. and Adams, N.G. (1980a). IAU Symposium, 87, 323.
Smith, D. and Adams, N.G. (]980b). Astrophys. J., 242, 424.
Smith, W.H., Snow, T.P. and York, D.G. (1977). Ap. J., 218, 124.
Snyder, L.E. (1972). MTP International Review of Science, Phys. Chem. Ser. 1 Vol. 3, Oxford, eds. A.D. Buckingham and D.A. Ramsay, p.193, Butterworth, London.

Snyder, L.E. and Buhl, D. (1971). Ap. J., 163, L47.
Snyder, L.E., Buhl, D., Zuckerman, B. and Palmer, P. (1969). Phys. Rev. Lett., 22, 679.
Snyder, L.E., Hollis, J.M., Ulich, B.L. (1976). Ap. J., 208, L91.
Solomon, P.M. and Klemperer, W. (1972). Ap. J., 178, 389.
Souza, S.P. and Lutz, B.L. (1977). Ap. J., 232, L175.
Suenram, R.D. and Lovas, F.J. (1978). J. Mol. Spectrosc., 72, 372.
Suenram, R.D. and Lovas, F.J. (1980). J.A.C.S. in press.
Thaddeus, P. (1972). Ann. Rev. of Astron. and Astrophys., 10, 305.
Thaddeus, P., Guélin, M. and Linke, R.A. (1981). Astrophys. J., 246, L41.
Tölle, F., Ungerechts, H., Walmsley, C.M., Winnewisser, G. and Churchwell, E. (1981). Astron. and Astrophys., 95, 143.
Townes, C.H. and Cheung, A.C. (1969). Astrophys. J., 157, L103.
Tucker, K.D., Kutner, M.L. and Thaddeus, P. (1974). Ap. J., 193, L115.
Turner, B.E. (1971). Ap. J., 163, L35.
Turner, B.E. (1974). Ap. J., 193, L83.
Turner, B.E. (1979). Bull. Am. Astron. Soc., 10, 627.
Turner, B.E. and Zuckerman, B. (1973). Ap. J., 187, L59.
Vishveshwara, S. and Pople, J.A. (1977). J.A.C.S., 99, 2422.
Waters, J.W., Gustincic, J.J., Kakar, R.N., Kuiper, T.B.H., Roscoe, H.K., Swanson, P.N., Rodriguez Kuiper, E.N., Kerr, A.R. and Thaddeus, P. (1980). Astrophys. J., 235, 57.
Watson, W.D. (1976). Rev. Mod. Phys., 48, 513.
Watson, W.D. (1977). Accounts of Chem. Res., 10, 221.
Watson, W.D. (1978). Ann. Rev. Astron. and Astrophys., 16, 585.
Weinreb, S., Barrett, A.H., Meeks, M.L. and Henry, J.C. (1963). Nature, 200, 829.
Wilson, R.W., Jefferts, K.B., Penzias, A.A. (1970). Ap. J., 161, L43.
Wilson, S. (1980). Chem. Rev., 80, 263.
Winnewisser, G., Churchwell, E. and Walmsley, C.M. (1979). Modern Aspects of Microwave Spectroscopy, ed. G. Chantry, p. 313, Academic Press, London.
Winnewisser, G., Churchwell, E. and Walmsley, C.M. (1979a). Astron. and Astrophys., 72, 215.
Winnewisser, G. and Walmsley, C.M. (1978). Astron. and Astrophys., 70, L37.
Woods, R.C., Dixon, T.A., Saykally, R.J. and Szanto, P.G. (1975). Phys. Rev. Lett., 35, 1269.
Wu, C.C. (1972). Ap. J., 178, 681.

PROBLEMS IN MODELLING INTERSTELLAR CHEMISTRY

D.A. Williams
Mathematics Department, UMIST, Manchester   M60 1QD, England

Abstract. This paper discusses solutions that are found by
modelling interstellar chemistry and argues that – though
mathematically correct – they may not always provide a satisfactory description of the astronomical objects. The problem
arises in the large array of data which are necessary and yet
which are poorly determined or unknown. The paper illustrates
the effects of uncertainties in these data. A modelling procedure which is potentially more reliable is outlined.

INTRODUCTION

It is a relatively straightforward task to construct a theoretical model of the chemistry in interstellar clouds. Using such a
model one might attempt to achieve several aims: the parameters of the
model may be varied to harmonize the theoretical results with molecular
column densities derived from observations; molecular abundances may be
predicted for molecules as yet unseen; parameters may be deduced from the
best-fit model apparently describing the interstellar cloud.

There are, however, very many parameter choices to be made in
such a calculation. Even fairly simple schemes involve large numbers of
gas-phase chemical reactions, but it is well known that only a small fraction of these reactions have measured rate coefficients, and of these the
temperature dependence is usually not well known. Reasonable estimates
may be made, but great uncertainties remain in the chemistry. Another
major problem in the chemistry arises with the interstellar grains. While
it is accepted that molecular hydrogen is formed on grain surfaces, it is
still controversial as to whether these surface reactions also contribute
other molecules (Williams 1980). As well as chemical parameters, there
are a large number of astrophysical parameters which need to be specified.
There are the parameters defining the local radiation field and cosmic
ray flux, there are parameters defining the cloud size and shape, density
and temperature. The elemental depletions relative to solar values must

be specified or determined. The effects on electromagnetic radiation of the grains, their scattering and absorbing properties, need to be known.

It is the purpose of this paper to point out that while the modelling procedure has been successful in describing the chemistry, it does not necessarily lead to unique descriptions of astronomical objects. The method, frequently adopted, of arbitrarily choosing some parameters to have fixed values and varying others to obtain a good fit with observational criteria carries no guarantee of finding the correct solution. It may be mathematically correct and physically sound but it need not be the solution which nature obtains in a free-parameter description. Although a unique mathematical solution does appear to exist for any particular choice of astrophysical or chemical parameters, changes in these parameters may produce inordinate changes in the solution for the chemistry. Nor do models which concentrate on certain aspects of the system to the exclusion of others, necessarily produce the same results. As an illustration of these points, Table 1 gives some density and temperature estimates for some recent models of the molecular cloud towards ζ Oph.

CHEMICAL UNCERTAINTIES

Table 2 lists some chemical routes, potentially of major importance, but which have uncertain rate coefficients. We illustrate

Table 1. Models of the ζ Oph Molecular Cloud

	$n(cm^{-3})$	$T(K)$
de Boer and Morton (1974)	272	56
Black and Dalgarno (1977)	500	110
	2500	22
Smith, et al. (1978)	120	56
Crutcher and Watson (1981)	200	65

Table 2. Some Chemical Processes with Uncertain
Rate Coefficients

$$C^+ + H_2 \rightarrow CH_2^+ + h\nu \qquad CH_3^+ + H_2 \rightarrow CH_5^+ + h\nu$$

$$H_3O^+ + e \rightarrow \begin{cases} H_2O + H \\ OH + 2H \end{cases} \qquad C + H_3^+ \rightarrow \begin{cases} CH_2^+ + H \\ CH^+ + H_2 \end{cases}$$

$$O + OH \rightarrow O_2 + H \qquad C + OH \rightarrow CO + H$$

$$CO + h\nu \rightarrow C + O \qquad NH + h\nu \rightarrow N + H$$

the type of difficulty encountered by discussing the dissociative recombination of H_3O^+ with electrons. The branching between the products OH
and H_2O is unknown, and conventionally it is assumed equally likely that
OH and H_2O is produced; a calculation by Herbst (1978) suggests the
OH:H_2O ratio is probably 10:1. Figure 1 illustrates the effect of varying
this ratio on the chemistry in a spherical $500M_\odot$ cloud of density 1000
H-nuclei cm^{-3} and temperature 30K. The results are column densities along
a diameter, normalized so that all values are 1.0 when the OH:H_2O ratio is
1. This single parameter is obviously important in the chemistry. Since
CO is unaffected by it, but OH directly by it, the CO:OH ratio
depends sensitively on this parameter. Figure 1 also shows what happens
when grains are permitted to contribute to the chemistry. The surface
chemistry adopted for these calculations is that of Pickles and Williams
(1977). It is clear that grains may modify the chemistry considerably.
Another major uncertainty in the chemistry is the rate of the radiative

Figure 1· the effect of varying the branching ratio in H_3O^++e
in a cloud of n = 1000cm^{-3}, T = 30K, M = $500M_\odot$. Column
densities along a diameter are normalized to unity for equal
branching, on the left the reaction yields OH, and on the
right H_2O. Solid lines – gas phase chemistry; dashed lines –
gas phase + surface chemistry.

Figure 2: n(OH)cm^{-3} as a function of position along a radius
of a cloud with n = 100cm^{-3}, M = $500M_\odot$, for various temperatures (K) as indicated; gas phase chemistry only.

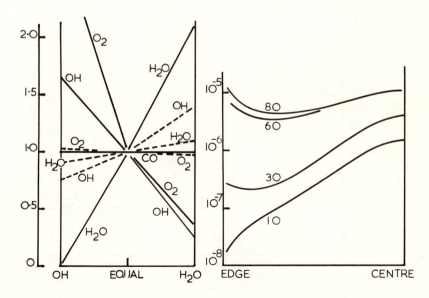

association: $C^+ + H_2 \rightarrow CH_2^+ + h\nu$. A value has been obtained by modelling the ζ Oph cloud (Black and Dalgarno 1977), but a range of acceptable values can be obtained in such a process by suitable adjustment of other parameters. The theoretical value for this rate coefficient is uncertain by a large factor (Herbst et al. 1977, Bates 1978).

ASTRONOMICAL UNCERTAINTIES

The initial choice to be made in modelling the chemistry is the definition of the geometry. Many models assume semi-infinite slab geometry, which has the disadvantage that very large masses may be implied for moderate optical depths. It seems safer to use spheres, except for thin sheets of gas. With spheres one must calculate column densities along a diameter: variations in local number densities may be more than an order of magnitude. No single point in the cloud is generally representative of the cloud; results given at a single point are meaningless. This point is well illustrated in Figure 2, which shows $n(OH) cm^{-3}$ in gas phase

Figure 3: the fractional column density $N(X)/N$ for various molecules X in clouds with $n = 1000 cm^{-3}$, $T = 30K$ as a function of cloud mass.

Figure 4: the effect of varying the cosmic ray ionization rate ζ on column densities along a diameter of a cloud with $n = 1000 cm^{-3}$, $T = 30K$, $M = 500M_0$; column densities are normalized to unity at $\zeta = 3 \times 10^{-18} s^{-1}$. Solid lines – gas phase chemistry; dashed lines – gas phase + surface chemistry.

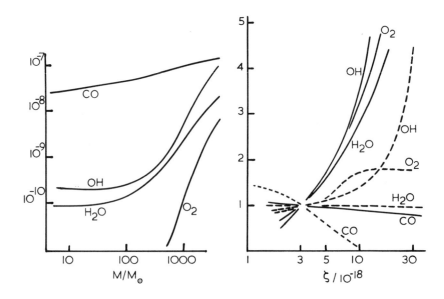

chemistry as a function of radius in a $500M_\odot$ cloud with n = $1000cm^{-3}$, the curves are labelled with the value of the temperature asssumed. The sensitivity with respect to temperature is high: in fact, it is a curious result that sensitivity is generally high with respect to astrophysical parameters near their "canonical" values. This fact is also illustrated in Figure 3 which shows the fractional column density N(X)/N for various molecules X as a function of cloud mass, in clouds of density n = $1000cm^{-3}$ and T = 30K.

The ionization rate, ζ, caused by cosmic rays, is also an important but unknown parameter. Figure 4 shows the effect of varying ζ on central densities in a medium density cloud; all the results have been normalized at unity for ζ = 3×10^{-18} s^{-1}. The behaviour is very sensitive to ζ but not all molecules behave in the same way. Results are also quite different, depending on whether grains contribute to the chemistry.

It has been realized that elemental depletions are also crucial parameters. The effect of changing the depletions are sometimes unexpected. Figure 5 exhibits the changes in column densities in a standard medium density cloud when the abundance of O and C is reduced from solar values (D=1) to one-third of this (D = $\frac{1}{3}$). It is not surprising that this change affects CO by an order of magnitude, but note that some species actually increase as the elemental abundance is reduced. Further unexpected effects occur when the O and C depletions vary independently. The level of metal depletion is also found to have important effects on the chemistry in dark clouds, for metal atoms and ions control the ionization and therefore the overall ion-molecule reaction rate.

Figure 5: the effect of changing abundances of O and C by a factor D on molecular column densities in a cloud of n = $1000cm^{-3}$, T = 30K; grain chemistry included.

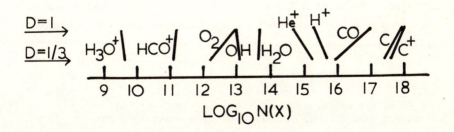

DISCUSSION

The overall sensitivity of results of this type of modelling
to parameter choice is very great. This sensitivity is often marked in
the region of parameter space most often selected. Severe confusion arises
in the interpretation of the models because the uncertainties in the
chemical description cannot be divorced from those of the astronomy.
Variations arising from some parameter changes are not intuitively obvious.
It is clear that we cannot expect unique descriptions of astronomical
objects to arise from modelling exercises which isolate different parts
of the chemistry or use different chemical and astrophysical parameters.

The situation will be improved as rate coefficients for
unknown reactions are accurately determined. It is particularly important
for the temperature dependence to be known for neutral exchanges and for
radiative associations. Some ion-molecule reactions also are temperature
dependent. As new observations help to define the nature of interstellar
grains, it should become possible to model accurately the contribution of
grains to the chemistry. Millar (elsewhere in this book) shows how surface reactions on oxide grains (Duley et al. 1979,1980) may modify chemistry in dark clouds.

An alternative approach is to attempt to model not individual
objects but a selected class of closely similar objects. This has the
advantage that at least some of the parameter choices are then common to
all members of the family, and the restricted choice for the remaining
parameters is less confusing. Some of the chemical ambiguity may be
removed in such studies. The possibility then arises of making unambiguous choices of astrophysical parameters. This method has been adopted
by Mann (1982). He has used the well-determined parameter E(B-V) to
classify clouds seen towards bright stars. For these clouds the chemistry
is radiation dominated and extinction is a significant parameter. This
general approach to the modelling problem may produce some success in
understanding these clouds, and may also be useful for other types of
clouds if they possess a suitable and well-determined classifying parameter.

REFERENCES

Bates, D.R. 1978 Proc.Roy.Soc. A360 1.
Black, J.H. and Dalgarno, A. 1977 Astrophys.J.Suppl. 34, 405.
deBoer, K.S. and Morton, D.C. 1974 Astr.Astrophys. 37, 305.

Crutcher, R.M. and Watson, W.D. 1981 Astrophys.J. 244, 855.
Duley, W.W., Millar, T.J. and Williams, D.A.
            1979 Astrophys.Sp.Sci. 65, 69.
            1980 Mon.Not.Roy.Astron.Soc. 192, 945.
Herbst, E. 1978 Astrophys.J. 222, 508.
Herbst, E., Schubert, J.G. and Certain, P.R. 1977 Astrophys.J. 213, 696.
Mann, A.P.C. 1982 in preparation.
Smith, A.M. Krishna Swamy, K.S. and Stecher, T.P. 1978 Astrophys.J. 220,
            138.
Pickles, J.P. and Williams, D.A. 1977 Astrophys.Sp.Sci. 52, 443.
Williams, D.A. 1980 J.dePhysique 41, C3-225.

ION-GRAIN COLLISIONS AS A SOURCE FOR INTERSTELLAR MOLECULES

T.J. Millar
Mathematics Department, U.M.I.S.T., P.O. Box 88, Manchester
M60 1QD

Abstract. In recent years it has been recognized that ionic
surface sites exist on a class of interstellar grains. This
raises the possibility that gas phase positively charged ions
can react with negatively charged surface ions to produce interstellar molecules. A brief discussion of the nature of the
surface sites is followed by an application to the chemistry
of diffuse and dense clouds. It is shown that ion-grain collisions can be the dominant formation mechanism for several molecular species.

INTRODUCTION

It has been known for many years that interstellar molecular
hydrogen formation occurs through the reaction of hydrogen atoms on the
surface of dust particles in the interstellar medium. Except under
special circumstances, the gas phase formation of H_2 is negligible.
Much of the early work on the formation of interstellar molecules was concerned with reactions on grain surfaces (Watson & Salpeter 1972a,b).
Since then, however, nearly all investigations, beginning with that of
Herbst & Klemperer (1973), have concentrated on gas phase schemes to
account for the increasingly detailed observations of interstellar molecules. This preponderance of studies of gas over grain reactions is due
to three major causes

    (i) the availability of accurate reaction rate coefficients,
      either theoretically or experimentally,

  (ii) the uncertain nature of the grain parameters, particularly
      their surface properties, and

(iii) the great success which ion-molecule reactions have had in
      explaining many of the observations, particularly those of
      HD and other deuterated species, and the observation of
      five interstellar molecular ions.

In recent years, however, there have been significant advances in our understanding of the nature and composition of interstellar grains. In particular, Duley and Millar have developed a model which argues that an important class of interstellar grains, namely those which provide the $\lambda 2200$ feature, is composed of diatomic oxide materials such as MgO, SiO and FeO. These particles must of necessity be small, probably less than 100 $\overset{\circ}{A}$ in radius, and provide a large surface area for the gas-grain interaction.

GRAIN CHEMISTRY IN DIFFUSE CLOUDS

In diffuse clouds atomic ions are the most important collision partners with oxide grains. Duley & Millar (1978) have investigated the reaction of such ions at surface sites on these oxide particles and have shown that such an interaction can account for the elemental depletions observed toward several stars including ζ Oph. Duley et al.(1978) showed that, under interstellar conditions, the most important reactive surface site on oxide grains is the $^{V}(OH^{-})_{s}^{-}$ site, a metal vacancy with a proton attached to a neighbouring O^{2-} ion. They also showed that ions, such as C^{+}, with a high ionization potential are not depleted onto the grain but can be returned to the gas in molecular form, as, for example, in CO. Millar et al. (1979) investigated the formation and destruction of H_2CO in diffuse clouds and showed that only C^{+} collisions with oxide grains proceed at a rate fast enough to account for its observed abundance. Their specific mechanism is the reaction of C^{+} at an $^{V}(OH^{-})_{s}^{-}$ site which has an associated weakly bound H-atom,

$$C^{+} + {}^{V}(OH^{-})_{s}^{-} \ldots H \rightarrow H_2CO + grain.$$

GRAIN CHEMISTRY IN DENSE CLOUDS

In clouds in which sulphur is mainly in neutral form, Duley & Millar (1979) argued that sulphur atoms replace oxygen atoms in the grain surface. This raises the possibility that sulphur molecules may be produced by ion collisions with grains in dense clouds. A preliminary study of this idea was presented by Duley et al. (1980) who showed that all of the then observed S-bearing molecules could be formed in this way. They also showed that the most important surface site in dense clouds is the SH^{-} site. It is ion collisions with these sites in particular that lead to molecule formation.

I have further developed the ideas formulated by Duley et al.
(1980) by studying a chemical network containing nearly 200 gas phase
reactions which is used to study the formation and destruction of 95 species. Almost 100 of these reactions are used to investigate the chemistry
of sulphur and silicon. In addition, over 20 grain reactions which produce S-bearing molecules can be included in the scheme. The grain reactions chosen include all of the most abundant ions in dense clouds. Thus
I include collisions of H^+, He^+, C^+, CH_3^+, CH_5^+, NH_3^+, NH_4^+, HCO^+ among
others. I discuss here the chemistry of some species for which grain
reactions make an important contribution.

(a) H_2S

It is difficult to form H_2S in the gas phase at an amount
adequate to explain the observations, $x(H_2S) \sim 3 \cdot 10^{-9}$ in Orion, because
the reaction $S^+(H_2, H)SH^+$ is endothermic. SH^+ can be formed through
$S(H_3^+, H_2)SH^+$ but the reaction of SH^+ with H_2 is also endothermic.
The radiative association $S^+(H_2, h\nu)SH_2^+$ is also very slow, so that the
net result is that only minor routes to H_2S exist. These routes predict $x(H_2S) \sim 10^{-11}$.

On grains, H_2S can be formed by a wide range of ions such as
H_3^+ and NH_4^+. For example,

$$H_3^+ + SH_s^- \rightarrow H_2S + H_2 + grain.$$

When grain reactions are added to the gas phase network with an efficiency
factor for molecule formation $\gamma_s = 0.05$, I find $x(H_2S) \sim 10^{-9}-2 \cdot 10^{-8}$
for $n = 10^3-10^6$ cm^{-3}, in good agreement with the observations.

(b) S_2

Diatomic sulphur forms through collision of S^+ with the
grains and I find $x(S_2) \sim 10^{-8}-10^{-9}$. Thus S_2 is predicted to be one of
the more abundant S-bearing species. Unfortunately it possesses no electric dipole moment although Liszt (1978) has searched for two of its magnetic dipole transitions without success. His upper limits are about a
factor of 10 greater than the above estimate. A method to detect S_2
indirectly would be to search for its protonated form, S_2H^+. The proton
affinity of S_2 is unknown but it could react with species such as
HCO^+, N_2H^+ and H_3^+. If so, then I estimate $x(S_2H^+) \sim 10^{-4}x(S_2)$.

(c) HNCS

This molecule has been detected in Sgr B2 with x(HNCS) ~ 2.5 10^{-11} (Frerking et al. 1979). A method of forming HNCS in gas phase reactions, analogous to the scheme proposed for HNCO by Huntress & Mitchell (1979) is via the radiative association reaction

$$NH_3^+ + CS \rightarrow H_3NCS^+ + h\nu$$

followed by $H_3NCS^+ + e \rightarrow HNCS + H_2$.

For similar association reaction rate coefficients, I find [HNCS]/[HNCO] $\simeq$ [CS]/[CO]. The observed ratios are [HNCS]/[HNCO] ~ $(0.3-1)10^{-2}$ and [CS]/[CO] ~ 10^{-4}, in disagreement with the theoretical estimate.

Reaction of HCNH$^+$ at the grain surface can lead to HNCS (and also to HCNS) and, with allowance for equal branching ratios to these two channels, I find x(HNCS) ~ $(0.2-1)10^{-11}$ for n = 10^4-10^6 cm^{-3}, in excellent agreement with the observations.

(d) CH$_3$SH

In a similar fashion, CH$_3$SH can be formed in the gas phase through the radiative association reaction

$$CH_3^+ + H_2S \rightarrow CH_3SH_2^+ + h\nu$$

followed by $CH_3SH_2^+ + e \rightarrow CH_3SH + H$.

I find [CH$_3$SH]/[CH$_3$OH] ~ [H$_2$S]/[H$_2$O]. Linke et al. (1979) find [CH$_3$SH]/[CH$_3$OH] ~ 7 10^{-3} in Sgr B2, whereas [H$_2$S]/[H$_2$O] is observed to be ~ $(0.2-2)10^{-5}$.

CH$_3^+$ collisions with grains can form CH$_3$SH. Linke et al. observe x(CH$_3$SH) ~ 1.5 10^{-10} while I predict x(CH$_3$SH) = $(0.6-2)$ 10^{-10} for n = 10^4-10^6 cm^{-3}.

(e) The HCS$^+$/CS Ratio

Thaddeus et al. (1981) have detected HCS$^+$ in several interstellar clouds and noted that it has an exceptionally high abundance relative to CS; N(HCS$^+$)/N(CS) ~ $(1-3)$ 10^{-2}. This is two orders of magnitude greater than that predicted by Mitchell et al. (1978). McAllister (1978) has predicted a value close to that observed but has erred in using values for [H$^+$] and [C$^+$] taken from Table 9 of Herbst & Klemperer

(1973). These values are larger than those found in more recent chemical models.

An investigation of the formation and destruction of CS, together with the observed HCS^+/CS ratio, leads to the conclusion that the rate coefficient, α_e, of the dissociative recombination reaction $HCS^+ + e \rightarrow CS + H$ must be less than 10^{-7} cm^3s^{-1} and is more likely to be closer to 10^{-8} cm^3s^{-1}. Such a value is much lower than those usually found experimentally for the dissociative recombination of polyatomic ions, typically around 10^{-6} cm^3s^{-1}.

If I choose $\alpha_e = 10^{-6}$ cm^3s^{-1}, then the chemical network predicts $[HCS^+]/[CS] \sim 10^{-4}$, whereas with $\alpha_e = 10^{-8}$ cm^3s^{-1}, the ratio is about $(5-10)\ 10^{-3}$, and agrees with that observed for $n \leq 10^5$ cm^{-3}.

SUMMARY

Ion collisions with interstellar oxide grains can play an important role in determining the chemical balance in interstellar clouds. In diffuse clouds the main effect of such collisions is to deplete the elements with ionization potentials close to the energy of the 2200 Å feature, i.e., 5.7 eV. However some simple oxygen-bearing molecules can be formed. In particular, a detailed model for the formation of H_2CO in diffuse clouds has been presented by Millar et al. (1979). In dense clouds, ion-grain collisions are likely to produce sulphur species. Although grain reactions cannot compete effectively with fast gas phase reactions, for example in the formation of SO, they may be the major source for species for which the gas phase routes are not straightforward, for example H_2S. The influence of grain reactions in the formation of several other S-bearing molecules has been discussed in detail by Millar (1981).

It is likely that the detailed chemistry in interstellar clouds depends on a complicated interplay between gas and grain reactions.

REFERENCES

Duley, W.W. & Millar, T.J. 1978, Ap.J., 220, 124.
Duley, W.W. & Millar, T.J. 1979, Ap.J.Letts., 233, L87.
Duley, W.W., Millar, T.J. & Williams, D.A. 1978, MNRAS, 185, 915.
Duley, W.W., Millar, T.J. & Williams, D.A. 1979, Ap.Sp.Sci., 65, 69.
Duley, W.W., Millar, T.J. & Williams, D.A. 1980, MNRAS, 192, 945.
Frerking, M.A., Linke, R.A. & Thaddeus, P. 1979, Ap.J.Letts., 243, L143.
Herbst, E. & Klemperer, W. 1973, Ap.J., 185, 505.

Huntress, W.T. & Mitchell, G.F. 1979, Ap.J., 231, 456.

Linke, R.A., Frerking, M.A. & Thaddeus, P. 1979, Ap.J.Letts., 243, L139.

Liszt, H.S. 1978, Ap.J., 219, 454.

McAllister, T. 1978, Ap.J., 225, 857.

Millar, T.J. 1981, MNRAS, to be published.

Millar, T.J., Duley, W.W. & Williams, D.A. 1979, MNRAS, 186, 685.

Mitchell, G.F., Ginsburg, J.L. & Kuntz, P.J. 1978, Ap.J.Suppl., 38, 39.

Thaddeus, P., Guélin, M. and Linke, R.A. 1981, Ap.J.Letts., 246, L41.

Watson, W.D. & Salpeter, E.E. 1972a, Ap.J., 174, 321.

Watson, W.D. & Salpeter, E.E. 1972b, Ap.J., 175, 659.

MILLIMETER AND SUBMILLIMETER LABORATORY SPECTROSCOPY: RECENT RESULTS OF ASTROPHYSICAL INTEREST

Frank C. De Lucia and Eric Herbst
Department of Physics, Duke University
Durham, North Carolina 27706

I. INTRODUCTION

Millimeter wave molecular line astronomy is rapidly expanding into the submillimeter spectral region (see for example, Phillips et al., 1980, Fetterman et al., 1981, Watson et al., 1980). Improvements in technology as well as increased awareness of the value of such observations have driven these advances. Even so, it is still a field that is technology limited and, in addition, severely restricted by the lack of atmospheric transparency. However, neither of these limitations is fundamental and technological innovation and the availability of space platforms should make possible even more rapid progress in the foreseeable future.

Many of the same technological limits exist for the laboratory spectroscopist. Nevertheless, a substantial body of work, mostly on small fundamental terrestrial molecules like H_2O, CO, etc., has been accumulated down to wavelengths of ~ 0.5 mm. (A bibliography of over 200 papers from the Duke Microwave Laboratory is available from the authors). However, many of the species of current interest to astronomers are not stable terrestrial molecules. The difficulty of producing significant amounts of unstable species in the laboratory combined with the technological problems have resulted in very little work on these astrophysically interesting molecules in the spectral region of interest.

A wide variety of reactive neutral species and molecular ions are predicted by current ion-molecule synthetic schemes (e.g., Prasad and Huntress 1980) to exist in interstellar clouds. Although these species are trace constituents of interstellar clouds, they are often present in sufficient abundance to produce observable emission spectra. A few have been identified by ingenious combination of molecular theory and astrophysical argument. Interesting recent examples are C_4H and C_3N (Guélin, Green, and Thaddeus, 1978, Guélin and Thaddeus, 1977). Nevertheless,

the list of observed interstellar species still shows a strong
bias towards those species with well established laboratory
spectra.

In this paper we will review our recent laboratory spectroscopic results on the transient neutral species CCH, NH_2, and
NaH, the molecular ions HCO^+, N_2H^+, and CO^+, and the light asymmetric internal rotor HOOH. We will show that the shorter millimeter and submillimeter spectral region has many advantages
even in the laboratory for this type of work. The requirements for theoretical models capable of the accurate spectral
predictions required for many astrophysical applications will
also be discussed.

II. <u>EXPERIMENTAL TECHNIQUES</u>

We have extensively discussed the millimeter and submillimeter spectroscopic techniques that have been used for the work
reported here (Helminger, De Lucia, and Gordy, 1970; De Lucia,
1976). Figure 1 shows the basic experiment. Microwave power

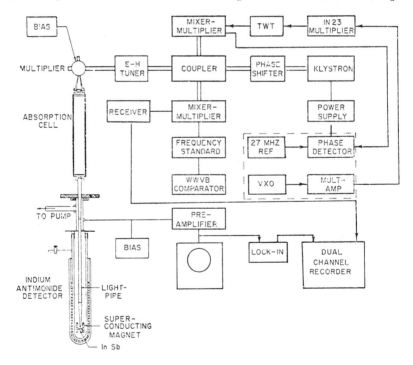

Figure 1. Block diagram of experiment.

in the 50 GHz region is produced by reflex klystrons. The
frequency of these klystrons is measured by comparison with
a 5 MHz oscillator, whose frequency is determined by a long
term phase comparison with WWVB. A similar comparison is used
to phase lock the klystrons when appropriate. The klystron
power is matched onto a crystal harmonic generator (King and
Gordy, 1953) and the harmonic power transmitted by quasi-optical techniques through the absorption cell. The energy is then
detected by an InSb detector operating at 1.4 K. Most of the
species reported here have been produced in sufficient abundance for observation in real time on an oscillascope. Others
have had their spectra recovered by use of either source or
Zeeman modulation and phase sensitive detection with time constants $\leq$ 1 sec. Digital signal averaging and computer processing have not been required for any of this work. The high
resolution, sensitivity, and relatively low cost of this technology have caused it to be widely adopted for spectroscopic
work in the shorter millimeter and submillimeter spectral regions.

It should be noted that the millimeter and submillimeter
spectral region is particularly advantageous for laboratory
studies of this nature. First, it provides directly the data
base required for extension of molecular radio astronomy to
shorter wavelength. In addition, absorption coefficients increase as approximately $\nu^2 \rightarrow \nu^3$ and very substantial gains in
detection sensitivity result, especially for those species
for which source modulation is appropriate. Furthermore, we
have shown (Clark and De Lucia, 1981) that the noise emissions
from discharge plasmas are insignificant at these wavelengths
and that plasma noise modulation of the probe signal decreases
as ν^2, reaching a negligible level at wavelengths shorter than
~1 mm. Finally, the quasi-optical techniques that allow the
probe signal to be focused through the chemical production region make possible the optimization of the production and lifetime of the transient species without consideration of the constraints imposed by microwave transmission requirements.

Many species that are abundant and important in the interstellar medium are difficult to produce in sufficient quantity
for spectroscopic studies because of their short life under
laboratory conditions. Although their laboratory production
is something of an art-form, knowledge of fundamental reaction
dynamics provides important guidance.

Figure 2 shows the cell that was used for the NaH work reported here (Sastry, Herbst, and De Lucia, 1981b). At the low

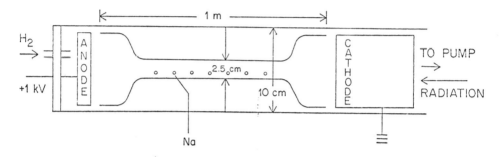

Figure 2.   Cell for production of NaH vapor.

pressures appropriate for gas phase microwave spectroscopy, NaH
decomposes upon heating rather than vaporizing. This is in contrast to the similar species LiD for which it was possible to
produce spectroscopically detectable quantities in a hot cell
experiment (Pearson and Gordy 1969). This decomposition has
made NaH an elusive species for gas phase absorption spectroscopists in all spectral regions. In the experiments described
in this paper, NaH was produced by reaction in a glow discharge
of hydrogen and sodium vapors. The absorption cell was a 2 m
length of 10 cm diameter pyrex pipe which contained continuously flowing hydrogen gas at ~0.2 Torr. Discharge electrodes
were placed at each end and between these electrodes was placed
a 1 m long, 2.5 cm diameter pyrex tube, flared to 9 cm at each
end. Approximately one gram of sodium metal was dispersed in
small pellets along the bottom of this tube. A discharge current of the order of one ampere flowed through the inner tube,
thereby heating and vaporizing the sodium pellets. At the end
of a several-hour run, a substantial fraction (~10%) of the
sodium had been converted into a grey-white powder, presumably
sodium hydride.

NH_2 (Charo, et al., 1981) was produced in the cell shown in Fig. 3.   This cell was a 10 cm diameter, 30 cm long Pyrex

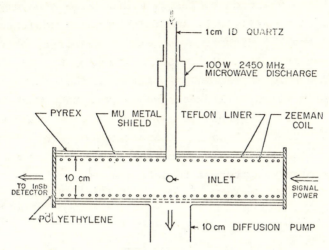

Figure 3.   Cell for the production of NH_2.

tube pumped by a 4" diffusion pump in a fast-flow configuration.   A magnetic modulation coil, a mu-metal shield, and an inner Teflon liner were located inside the cell.   The cell was modulated at 1300 Hz, and the signal recovered at 2600 Hz by means of a lock-in amplifier with a 1 s time constant.   The NH_2 was produced by the reaction of the products of a 2450 MHz microwave discharge of H_2O with N_2H_4 (Hills et al. 1976).

The nitrogen cooled, glow discharge cell shown in Fig. 4 was used to produce CCH (Sastry, et al., 1981a) as well as the molecular ions reported here.   Ever since the astronomical dis-

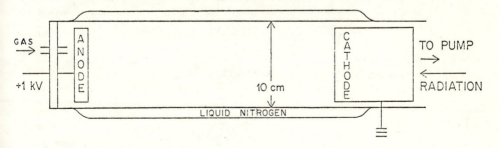

Figure 4.   Liquid nitrogen cooled discharge cell.

covery of this species by Tucker, Kutner, and Thaddeus (1974),
CCH had been sought by laboratory spectroscopists. Our approach
was to adopt the nitrogen cooled glow discharge commonly used
for ions (Woods et al., 1975) and to choose non-condensable (at
77 K) precursors. A discharge current on the order of 0.3 A
was maintained in a 2 : 1 : 0.1 mixture of He, CH_4, and CO
at a total pressure of 25 mtorr. Although the stronger signals
were marginally video, all measurements were made with phase locke
klystrons and lock-in signal recovery.   Figure 5 shows the
N=2→3, J=5/2→7/2, F=2→3 and 3→4 transitions of CCH. We expect
this to be a general technique for the production of many small
transient species of interest to astronomers.

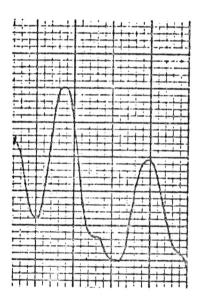

Figure 5.   CCH transition at 262 GHz.

HCO^+ and DCO^+ (Sastry, Herbst, and De Lucia, 1981c) were produced in the same cell with the total pressure optimized at ~40 mtor with equimolar amounts of CO and $H_2(D_2)$. However, discharge currents of 100-500 ma reduced the pressure on the pump end of the cell to ~20 mtorr, presumably by production of condensable species from the CO and $H_2(D_2)$. Under the conditions of our experiment, the HCO^+ spectra were quite strong, absorbing about one percent of the power in the region around 300 GHz.

For CO^+ (Sastry, et al., 1981d) a total pressure of 30 mtorr with a CO/He mole ratio of 1:2 was used. The observed signal-to-noise ratios were comparable with those obtained in this laboratory for HCO^+ and DCO^+.

The species HN_2^+ and DN_2^+ were produced similarly (Sastry, et al. 1981e). The total pressure was optimized at ~80 mtorr with an N_2/H_2 or N_2/D_2 molar ratio of 1 : 6. Discharge currents of 350-500 ma reduced the pressure on the pump end of the cell to ~60 mtorr. The observed signal-to-noise ratios were comparable with, but somewhat smaller than those measured by us for HCO^+ and DCO^+.

HOOH (Helminger, Bowman, and De Lucia, 1981; Bowman, De Lucia, and Helminger, 1981) was slowly pumped through a 3m long, 2.5cm diameter metal cell with teflon windows. Since hydrogen peroxide reacts rapidly with metal surfaces, the inside of this cell was coated with plastic krylon spray.

III.   ROLE OF THEORY

Molecular radio astronomy has its basis in laboratory measurements. These measurements not only provide the accurate rest frequencies necessary for the interpretation of astronomical data, but also make possible the proper assignment of lines from the ever increasing background density that has resulted from increased detector sensitivity.

Since it is not practical to observe all possible transitions of potential astrophysical importance, it is important to consider the accuracy with which theoretical spectral maps can be constructed from a limited number of measurements and when such a calculation is appropriate.

Linear and symmetric top molecules in well behaved electronic states (i.e. those sufficiently removed from other electronic states that their rotational spectra can be treated via "effective" constants) often have closed form expressions for their rotational energy. More importantly, because there is a systematic relation between transition frequency and the rotational quantum number N (in the absence of electronic spin, N = J)

$$\nu \sim 2B(N+1),$$

the transitions at high N, with potentially large distortions, ordinarily do not fall in the spectral region of interest. The situation is not as sanguine for light asymmetric rotors because the reduced symmetry of the molecules allows transitions between closely spaced, but highly excited rotational levels. For the linear species reported in this paper (ignoring hyperfine interactions)

$$^1\Sigma \text{ electronic state} \quad : \quad \nu = 2B(J+1)-4D(J+1)^3 \qquad (1)$$

$$^2\Sigma \text{ electronic state} \quad : \quad \nu = 2B(N+1)-4D(N+1)^3 \pm \gamma/2 \qquad (2)$$

(In the latter case are included only the strong $\Delta N = \Delta J = 1$ transitions).

The linear species NaH, HCO^+, and HN_2^+ possess ground electronic states of $^1\Sigma$ symmetry and rotational spectra that, neglecting electric quadrupole moment interactions (important for HN_2^+), obey Eq. 1.

The molecular ion CO^+ possesses a ground electronic state of $^2\Sigma^+$ symmetry. The three quantum numbers utilized are $\bar{N}$, the rigid body angular momentum quantum number, $\bar{S}=1/2$, the electron spin angular momentum quantum number, and $\bar{J}=\bar{N}\pm1/2$, the total angular momentum quantum number. The spectra can be fit to microwave accuracy with three molecular parameters--B, the rotation constant, D, the centrifugal distortion constant, and γ, the spin rotation constant, via Eq. 2.

The radical CCH also possesses a ground electronic state of $^2\Sigma$ symmetry but has the additional complication of hyperfine interactions. The angular momentum coupling scheme (Hund's case $b_{\beta J}$) most appropriate for its description is $\bar{N} + \bar{S} = \bar{J}$, $\bar{J} + \bar{I} = \bar{F}$ where $\bar{N}$ is the angular momentum of the inertial frame, S is the electron spin, $\bar{I}$ is the nuclear spin of the hydrogen, and where $\bar{J}$ and $\bar{F}$ are defined by the coupling scheme. The coupling of the real molecule deviates from this pure coupling scheme because the hyperfine interaction is not small compared to the spin-doubling. Thus matrix elements off diagonal in $\bar{J}$ must be included. Although this requires a small matrix diagonalization, reliable calculations of transition frequencies well into the submillimeter spectral region can still be expected.

The remaining species reported here are light asymmetric rotors. We have extensively discussed in the literature model building and calculation of synthetic spectra for molecules of this type (Helminger, Cook, and De Lucia, 1971; Cook, De Lucia, Helminger, 1972). Briefly, Watson's (1966) power series Hamiltonian is adopted as the theoretical model for molecules without internal rotation or electronic effects. We have shown that this approach is capable of fitting large experimental data sets to within experimental uncertainty (~0.1 MHz), and that accurate calculation of unobserved transition frequencies is possible <u>if the transitions to be calculated are similar to those included in the data base of the analysis</u> (in linear and symmetric top molecules this requirement is always met). If they are not, and especially if the molecule is light and requires many distortion constants in its analysis, the correlation among the higher order constants makes calculation of such transition frequencies unreliable.

NH_2 is an example of a light asymmetric rotor with substantial centrifugal distortion, with the additional complications of electronic effects and nuclear hyperfine structure. The ground electronic state of NH_2 possesses $^2B_1$ symmetry (Dressler and Ramsay 1957). In addition to electron-spin angular momentum, three nuclear spins must be considered. An appropriate coupling scheme is $\bar{N}+\bar{S}=\bar{J}$, $\bar{I}_{H_1}+\bar{I}_{H_2}=\bar{I}_H$, $\bar{J}+\bar{I}_H=\bar{F}_H$, and $\bar{F}_H+\bar{I}_N=\bar{F}$, where $\bar{N}$ is the rigid-body angular momentum; $\bar{S}$, the electron-spin angular momentum; $\bar{I}_{H_1}$, $\bar{I}_{H_2}$, and $\bar{I}_N$, the spin angular momenta of the respective nuclei; and $\bar{J}$, $\bar{I}_H$, $\bar{F}_H$, and $\bar{F}$ are defined by the coupling scheme. The NH_2 molecule has two identical hydrogen nuclei and, consequently, possesses ortho $(I_H=1)$ and para $(\bar{I}_H=0)$ states. In the 3_{13} and 2_{20} levels, $\bar{I}_H=1$, and for the 1_{10} and 1_{01} levels, $\bar{I}_H=0$.

Unfortunately, the number of NH_2 transitions that fall in the millimeter and submillimeter spectral region are far too few to allow calculation of the rotational spectral constants that characterize the species. Fortunately, its spectrum is so sparse that the transitions of astrophysical importance can be observed directly (see below), and the difficulty associated with modeling is not a serious problem.

HOOH is a light asymmetric rotor with additional spectral complexity due to its internal rotation. Unlike NH_2 it has a rich spectrum with several hundred strong rotational transitions in the region of interest. Thus sufficient data for a theoretical analysis can be obtained. We have discussed elsewhere (Helminger, Bowman, and De Lucia, 1981) a theoretical model that allows the analysis of this molecule to within experimental uncertainty. While we report the direct measurement of many of the lines of interest, this theoretical model is important for confirmation of assignments and calculation of unobserved transitions.

IV.   RESULTS AND ANALYSES

Tables I through VII show the observed rotational transitions of CCH (Sastry, et al., 1981a); NaH (Sastry, Herbst, and De Lucia, 1981b); NH_2 (Charo, et al., 1981); HCO^+ (Sastry, Herbst, and De Lucia, 1981c); CO^+ (Sastry, et al., 1981d); and HN_2^+ (Sastry et al., 1981e) respectively. We have also reported an extensive

TABLE 1.   Rotational Transitions of CCH (MHz).

N"	N'	J"	J'	F"	F'	Observed	Obs.-Cal.
1 → 2		3/2 → 5/2		2 → 3		174663.179	-0.012
				1 → 2		174667.642	+0.035
		1/2 → 3/2		1 → 2		174721.765	-0.013
				0 → 1		174728.048	-0.010
2 → 3		5/2 → 7/2		3 → 4		262004.260	-0.038
				2 → 3		262006.482	+0.015
		3/2 → 5/2		2 → 3		262064.986	+0.010
				1 → 2		262067.469	+0.013
3 → 4		7/2 → 9/2		a		349338.103	+0.260
		5/2 → 7/2		a		349400.612	+0.116

[a] Hyperfine structure only partially resolved, not included in fit.

TABLE II.   Rotational Transitions of Sodium Hydride (MHz)

	J"	J'	Frequency Observed
NaD	0 → 1		151765.40[a] (15)
	1 → 2		303463.80 (20)
	2 → 3		455028.14 (40)
NaH	0 → 1		289864.35 (20)

[a] Calculated from fitted constants.

TABLE III. Components of the $2_{20} \to 3_{13}$ Transitions of NH_2 (MHz).

F''_H	F''	$\to$	F'_H	F'	Frequency	Relative Intensity
					A. $J'' = 3/2 \to J' = 5/2$	
3/2	5/2		5/2	7/2	229405.86 ± 0.07^{a}	5
5/2	7/2		7/2	9/2	229409.87 ± 0.03	8
5/2	5/2		7/2	7/2	229474.94 ± 0.06	6
3/2	5/2		7/2	7/2	229494.38 ± 0.03	b
5/2	3/2		7/2	5/2	229528.04 ± 0.06	4
1/2	3/2		5/2	5/2	229481.40 ± 0.06	b
					B. $J'' = 5/2 \to J' = 7/2$	
$\left\{\begin{matrix}7/2\\5/2\end{matrix}\right.$	$\left.\begin{matrix}5/2\\3/2\end{matrix}\right\}$		$\left\{\begin{matrix}9/2\\7/2\end{matrix}\right.$	$\left.\begin{matrix}7/2\\5/2\end{matrix}\right\}$	241533.95 ± 0.10^{c}	$\left\{\begin{matrix}6\\4\end{matrix}\right\}$
3/2	1/2		5/2	3/2	241538.80 ± 0.50	2
$\left\{\begin{matrix}7/2\\5/2\end{matrix}\right.$	$\left.\begin{matrix}7/2\\5/2\end{matrix}\right\}$		$\left\{\begin{matrix}9/2\\7/2\end{matrix}\right.$	$\left.\begin{matrix}9/2\\7/2\end{matrix}\right\}$	241557.68 ± 0.06^{c}	$\left\{\begin{matrix}8\\7\end{matrix}\right\}$
3/2	3/2		5/2	5/2	241561.83 ± 0.07	4
$\left\{\begin{matrix}5/2\\7/2\end{matrix}\right.$	$\left.\begin{matrix}7/2\\9/2\end{matrix}\right\}$		$\left\{\begin{matrix}7/2\\9/2\end{matrix}\right.$	$\left.\begin{matrix}9/2\\11/2\end{matrix}\right\}$	241591.33 ± 0.09^{c}	$\left\{\begin{matrix}8\\10\end{matrix}\right\}$
3/2	5/2		5/2	7/2	241596.00 ± 0.26	6

[a] Standard deviation from at least five averages of forward and reverse sweeps.

[b] Relative intensity of 0 with a first-order calculation.

[c] Absorption maximum for unresolved pair.

TABLE IV.   Components of the $1_{01} \rightarrow 1_{10}$ Transition of NH_2 (MHz).

F" → F'	Frequency	Relative Intensity
	A. J" = 1/2 → J' = 3/2	
3/2 → 5/2	$461,465.03 \pm 0.16$[a]	2
	B. J" = 3/2 → J' = 3/2	
5/2 → 3/2	$462,425.42 \pm 0.06$	2
5/2 → 5/2	$462,433.51 \pm 0.06$	10
3/2 → 1/2	$462,449.04 \pm 0.10$	2
3/2 → 3/2	$462,455.58 \pm 0.06$	4
1/2 → 1/2	$462,467.02 \pm 0.07$	2
	C. J" = 1/2 → J' = 1/2	
3/2 → 3/2	$469,440.62 \pm 0.06$	4
1/2 → 3/2	$469,383.37 \pm 0.06$	3
3/2 → 1/2	$469,366.12 \pm 0.06$	3

[a]Standard deviation from at least five averages of forward and reverse sweeps.

TABLE V.   Rotational Transitions of HCO^+ and DCO^+ (MHz).

J" J'	HCO^+ Observed	Obs.-Calc.	DCO^+ Observed	Obs.-Calc.
1 → 2	178375.065(50)[a]	0.022	144077.319(50)[a]	0.003
2 → 3	267557.619(10)[a,b]	-0.001	216112.604(50)[a]	-0.018
3 → 4	356734.288(50)	0.022	288143.911(50)[a]	0.006
4 → 5	445902.996(50)	-0.005	360169.881(100)[a]	0.056
5 → 6	---	---	432189.033(50)	-0.008

[a]Also detected by Bogey, Demuynck, and Destombes (1981).

[b]Huggins, et al. (1979) have obtained an astronomical frequency 267557.20$\overline{(46)}$ MHz.

TABLE VI. Rotational Transitions of CO^+ (MHz).

N"	N'	J"		J'	FREQUENCY
1 → 2		1/2	→	3/2	235789.641(30)
1 → 2		3/2	→	5/2	236062.553(20)
1 → 2		3/2	→	3/2	235380.046(150)[a]
2 → 3		3/2	→	5/2	353741.262(100)
2 → 3		5/2	→	7/2	354014.247(60)
3 → 4		5/2	→	7/2	471679.213(120)
3 → 4		7/2	→	9/2	471952.343(100)

[a]Not included in least squares fit because of severe Zeeman broadening.

Table VII. Rotational Transitions of HN_2^+ and DN_2^+ (MHz).

J"		J'	Observed[a]		Corrected[b]	Corrected-Calculated
				HN_2^+		
1	→	2	186344.874(100)		186344.771	0.118
2	→	3	279511.671(50)[c]		279511.701	-0.028
3	→	4	372672.497(50)		372672.509	0.003
4	→	5	465824.941(250)		465824.947	0.064
				DN_2^+		
1	→	2	154217.199(50)		154217.096	0.174
2	→	3	231321.635(50)		231321.665	-0.069
3	→	4	308422.210(50)		308422.222	0.055
4	→	5	385516.756(100)		385516.762	-0.001
5	→	6	462603.931(200)		462603.932	-0.130

[a]Includes shift to cancel Doppler effect caused by ion drift velocity.

[b]Corrected to remove quadrupolar contribution.

[c]Erickson et al.(1981) have obtained an astronomical frequency of 279513 MHz in 10 sources.

analysis of the excited states of NaH and NaD through v=3 (Sastry, Herbst, and De Lucia, 1981f). This work was possible because our production technique produces vibrationally hot sodium hydride. Except for NH_2 these are linear species that may be described by the simple closed form expressions of Eqns. 1 and 2. Table VIII collects together the spectral constants calculated via these equations from the data in these tables. Higher J(N) transitions

TABLE VIII. Spectral constants for Linear Species.

Molecule	Type	B_o (MHz)	D_o (kHz)	γ_o (MHz)
CO^+	$^2\Sigma$	58983.040(12)	189.6(5)	273.01(5)
CCH	$^2\Sigma$	43674.515(6)	105.3(4)	−62.653(19)
HN_2^+	$^1\Sigma$	46586.863(15)	87.50(53)	---
DN_2^+	$^1\Sigma$	38554.717(14)	60.81(37)	---
HCO^+	$^1\Sigma$	44594.420(2)	82.39(7)	---
DCO^+	$^1\Sigma$	36019.776(3)	55.87(5)	---
NaH	$^1\Sigma$	144952.7(3)	10280.(30)	---
NaD	$^1\Sigma$	75888.29(8)	2793.(8)	---

calculated from these data should be reliable.

Unlike the rest of these species that have only a few transitions in this spectral region, HOOH has many. Indeed, more than 250 have been observed. Although these transitions are too extensive to list here, they may be found in the literature (Helminger, Bowman, and De Lucia, 1981, Bowman, Helminger, and De Lucia, 1981). Figure 6 shows the observed transitions in the form of a FORTRAT diagram. Table IX shows the spectral constants that result from this work.

V. CONCLUSIONS

The recent results shown here demonstrate that millimeter and submillimeter microwave spectroscopy is a sensitive and flexible means of studying short-lived species in the laboratory.

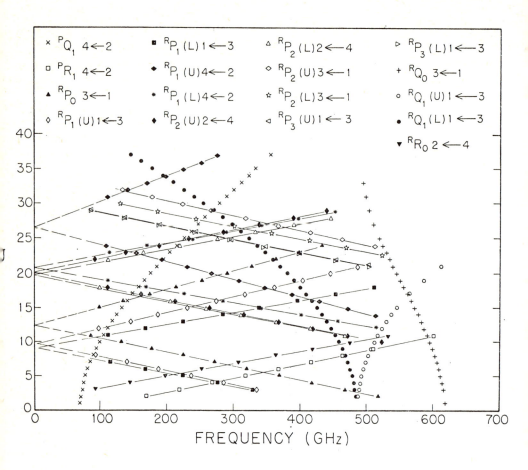

Figure 6.   FORTRAT diagram for HOOH.

Table IX. Rotational Constants of HOOH (MHz).

	$\tau = 1,2$		$\tau = 3,4$	
	value	σ	value	σ
A	301 874.205	0.037	301 586.074	0.075
B	26 212.439	0.046	26 155.639	0.036
C	25 098.604	0.046	25 186.461	0.036
$\Delta_J(\cdot 10^0)$	0.105292	0.0000029	0.0996169	0.000014
$\Delta_{JK}(\cdot 10^{-1})$	0.112180	0.000026	0.115275	0.000051
$\Delta_K(\cdot 10^{-2})$	0.120340	0.00012	0.118558	0.00023
$\delta_J(\cdot 10^3)$	-0.258456	0.0014	0.686780	0.0012
$\delta_K(\cdot 10^{-1})$	0.978631	0.0027	0.638326	0.00021
$H_{JK}(\cdot 10^3)$	0.506307	0.0097	0.0594176	0.0013
$H_{KJ}(\cdot 10^2)$	-0.161735	0.0087	-0.0506399	0.0073
$H_K(\cdot 10^2)$	0.185268	0.097	---	
$h_J(\cdot 10^7)$	0.304918	0.022	-0.515225	0.011
$h_{JK}(\cdot 10^3)$	0.172789	0.00090	0.129727	0.0011
$h_K(\cdot 10^1)$	0.440464	0.038	-0.631712	0.0028
$L_{JJK}(\cdot 10^7)$	-0.159961	0.0045	-0.105713	0.0071
$L_{JK}(\cdot 10^5)$	-0.933110	0.051	0.829787	0.0077
$L_{KKJ}(\cdot 10^4)$	0.218515	0.038	---	
rms	0.126		0.112	
number of data points	119		141	
W		342 881.62		

These and similar studies, when combined with the appropriate
theoretical approach, are capable of providing accurate spectral
maps over wide spectral regions for most species of astrophysical importance.

It is important to note that this work has been accomplished
in a relatively short period of time and by use of only our
basic spectroscopic technology. Digital signal processing (e.g.
integration), more sophisticated cell design, and improved
chemical diagnostics all are relatively straightforward escala-
tions. Thus substantial progress is still to be expected.

VI.   ACKNOWLEDGMENTS

We would like to acknowledge the collaboration of Wayne Bowman
Arthur Charo, Paul Helminger, and K.V.L.N. Sastry in these experi-
ments. We also acknowledge the support of NASA through grant
NAGW-189.

BIBLIOGRAPHY

M. Bogey, C. Demuynck, J.L. Destombes, private communication.

W. Bowman, F.C. De Lucia, and P. Helminger, J. Mol. Spectrosc.
87, 571 (1981).

A. Charo, K.V.L.N. Sastry, E. Herbst, and F. C. De Lucia, Ap.
J. Letters 244, L111 (1981).

W.W. Clark, III and F.C. De Lucia, J. Chem. Phys. 74, 3139 (1981).

R.L. Cook, F.C. De Lucia, and P. Helminger, J. Mol. Spectrosc.
41, 123 (1972).

F.C. De Lucia, in Molecular Spectroscopy, Modern Research Vol. II,
(Academic Press, N.Y.)(1976).

K. Dressler and D.A. Ramsey, J. Chem. Phys. 27, 971 (1957).

N. Erickson, J.H. Davis, N.J. Evans, II, R.B. Loren, L. Mundy,
W.L.Petters, III, M. Scholter and P.A.Vanden Bout, private
communication.

H.R. Fetterman, G.A. Koepf, P.F. Goldsmith, B.J. Clifton,
D.Buhl, N.R. Erickson, D.D. Peck, and P.E. Tannewald, Science 211,
580 (1981).

M.Guélin, S. Green, and P. Thaddeus, Ap. J. Letters 224, L27 (1978

M. Guélin and P. Thaddeus, Ap. J. Letters 212, L81 (1977).

P. Helminger, R.C. Cook, and F.C. De Lucia, J.Mol.Spectrosc. 40, 125 (1971).

P. Helminger, W.C. Bowman, and F.C. De Lucia, J. Mol. Spectrosc. 85, 120 (1981).

P. Helminger, F.C. De Lucia, and W. Gordy, Phys. Rev. Letters 25, 1397 (1970).

G.W. Hills, J.M. Cook, R.F. Curl, Jr., and F.K. Tittel, J. Chem. Phys. 65, 823 (1976).

P.J. Huggins, T.G. Phillips, G. Neugebauer, M.W. Werner, P.G. Wannier, and D. Ennis, Astrophys. J. 227, 441 (1979).

W.C. King and W. Gordy, Phys. Rev. 90, 319 (1953).

E.F. Pearson and W. Gordy, Phys. Rev. 177, 59 (1969).

T.G. Phillips, P.J. Huggins, T.B.H. Kuiper, and R.E. Miller, Ap. J. Letters 238, L103 (1980).

S.S. Prasad and W.T. Huntress, Jr., Astrophys. J. Supp. Ser. 43, 1 (1980).

K.V.L.N. Sastry, P. Helminger, E. Herbst, and F.C. De Lucia, submitted for publication (1981a).

K.V.L.N. Sastry, E. Herbst, and F.C. De Lucia, Ap. J. Letters, to appear (1981b).

K.V.L.N. Sastry, E. Herbst, and F.C. De Lucia, J. Chem. Phys. (Letters), to appear, (1981c).

K.V.L.N. Sastry, P. Helminger, E. Herbst, and F.C. De Lucia, Ap. J. Letters, to appear (1981d).

K.V.L.N. Sastry, P. Helminger, E. Herbst, and F.C. De Lucia, submitted for publication (1981e).

K.V.L.N. Sastry, E. Herbst, and F.C. De Lucia, J. Chem. Phys. to appear (1981f).

K.D. Tucker, M.L. Kutner, and P. Thaddeus, Ap. J. Letters 193, L115 (1974).

J.K.G. Watson, J. Chem. Phys. 45, 1360 (1966).

D.M. Watson, J.W.V. Storey, C.H. Townes, E.E. Haller, and W.L. Hansen, Ap. J. Letters 239, L129 (1980).

R.C. Woods, T.A. Dixon, R.J. Saykally, and P.G. Szanto, Phys. Rev. Letters 35, 1269 (1975).

LIMITS ON THE |D|:|H| RATIO IN THE INTERSTELLAR MEDIUM FROM MOLECULAR OBSERVATIONS.

Robin L. Frost, John E. Beckman, Graeme D. Watt, Glenn J. White and J. Peter Phillips.
Department of Physics, Queen Mary College, Mile End Road, London E1 4NS.

Abstract. The abundance ratio of |D|:|H| in the interstellar medium can be derived either more directly from the UV absorption measurements of HD and H_2 in the spectra of early-type stars, or less directly from millimetre-wave emission in the rotational lines of a variety of molecules. The latter method is applicable over a much larger portion of the galaxy, but is subject to the uncertainty inherent in chemical fractionation. It is shown here how the basic assumptions of temperature distribution and geometry within a single source can cause the derived ratio |HDO|:|H_2O| to vary between 5×10^{-2} and 2.5×10^{-3}, a comparable range to that found with three different molecules |DCN|:|HCN|, |DNC|:|HNC| and |DCO^+|:|HCO^+| in the same source. This result should lead to caution in the interpretation of such abundance ratios and of their gradients within the galaxy.

Introduction

Evaluation of the |D|:|H| ratio over as wide a range of physical conditions as possible is important in order to test first qualitatively and then quantitatively the now widely accepted idea that all the deuterium we can observe was produced within the primaeval explosion (Wagoner, 1973). Interstellar deuterium can be detected in one of two ways, either using the isotopically shifted absorption lines in the spectra of appropriately reddened O and B stars, or using the rotational emission spectra of deuterated molecules. The former method is at present more precise, in that the |HD|:|H_2| ratio is found directly, but for a very restricted set of interstellar lines of sight out to 2 kpc from the sun. The second and more recent method is available for a much larger portion of the galaxy, including the galactic centre, but suffers from the interpretative uncertainties inherent in chemical fractionation.

As an example of this difficulty, Table 1 gives a
selection of recently measured values of molecular abundance
ratios in a single source, the Kleinmann-Low source in
the Orion Nebula.

There is a spread of a factor forty in these
values, and collectively they differ markedly from the
values in the range $\sim 10^{-5}$ obtained using the Lyα method
by a variety of UV observers.

Model dependence in the molecular determination of $|H|:|D|$.

In order to show that the range of abundance
differences illustrated in Table 1 may be due to model-
dependent considerations as well as to real differences
in chemical behaviour, we will use the example of $|H_2O|:|HDO|$.
There are now available two observed lines of HDO,
the $1(1,0) \rightarrow 1(1,1)$ rotational transition at 80.6 GHz
observed by Turner et al. (1975), and the $2(1,1) \rightarrow 2(1,2)$
line at 272 GHz reported elsewhere by the present authors
(Beckman et al. 1982). One may use the intensities of
these lines in conjunction to compute the HDO column
density, and this may be compared with the H_2O column
density, derived from the $3(1,3) \rightarrow 2(2,0)$ transition at
183 GHz by Waters et al. (1980).

In order to compute the HDO column density
we first assume that the portion of the cloud giving
rise to the $2(1,1) \rightarrow 2(1,2)$ transition is isothermal,
with a temperature T_{KIN} (effects of departure from this
assumption are shown below). We can then use a Boltzmann

Table 1

Measured ratio	Value	Reference
$|DCN|:|HCN|$	1.17×10^{-3}	Penzias 1979
$|DNC|:|HNC|$	5×10^{-2}	Snell & Wootten, 1979
$|DCO^+|:|HCO^+|$	2.3×10^{-3}	Penzias 1979
$|HDO|:|H_2O|$	$<3 \times 10^{-3}$	Rodriguez-Kuiper et al. 1978

approximation to write down

$$\frac{\tau_{HDO}}{N_{HDO}} = \frac{f_u c^3 A_{22}}{8\nu^3 \Delta\nu} \quad \exp(h\nu/kT_{KIN}) - 1 \tag{1}$$

with $\qquad f_u = \frac{N_u}{N} = g_u \exp(E_u/kT_{KIN}) \tag{2}$

where ν is the rest frequency of the emission line, $\Delta\nu$ the half-width at half-intensity, A_{22} is the Einstein coefficient of the $2(1,1) \rightarrow 2(1,2)$ transition, g_u the degeneracy of the upper level, and E_u the energy level above the ground state of the upper level. The plot in Fig. 1 comes from calculations using (1) varying T_{KIN} between 5K and 200K, as well as the equivalent calculations for H_2O.

To calculate the abundance ratio r defined by

$$r = \frac{N_{HDO}}{N_{H_2O}} \tag{3}$$

we can organize the algebra in the form

$$r \cdot \left\{ \frac{\tau_{H_2O}}{\tau_{HDO}} \right\} = \left\{ \frac{\tau_{H_2O}}{N_{H_2O}} \right\} \cdot \left\{ \frac{N_{HDO}}{\tau_{HDO}} \right\} \cdot \tag{4}$$

An initial method for r then invokes the inequality

$$T_b \lesssim T_{KIN} \tau \tag{5}$$

which goes to its limiting form of an equation as $\tau \rightarrow 0$. One can derive T_{KIN} from the line centres of clearly optically thick lines, of which there are several in the CO ladder. A reasonably good value (e.g., Phillips et al. 1977) is $T_{KIN} = 70K$. With the value of the corrected antenna temperature T_b for HDO as 1K, we find $\tau_{HDO} = 0.015$, which is optically thin, justifying the use of the asymptotic form in (5). We can then, using Fig. 1, derive a value for N_{HDO} of $4.76 \times 10^{13} m^{-2}$. Using the equivalent data for H_2O we find $N_{H_2O} \lesssim 2.15 \times 10^{15} m^{-2}$, i.e.

$$|HDO| : |H_2O| < 2.2 \times 10^{-2} \qquad\qquad I$$

This has to be an upper limit, because an
optically thin assumption for H_2O is not supported by
the measured antenna temperature for the $3(1,3) \rightarrow 2(2.0)$
transition of H_2O, which is 15K.

An improvement can be made by assuming that HDO
is indeed optically thin, but the H_2O transition is
optically thick. We can then derive the expression

$$\tau_{HDO} = -\ln\left\{1 - \frac{T_b(HDO)}{T_b(H_2O)} \cdot \frac{T_{ex}(H_2O)}{T_{ex}(HDO)}\right\} \qquad (6)$$

where T_b for each molecule is the observed antenna
temperature, and T_{ex} the excitation temperature in
the region where the line is formed. It is then necessary
to assume that beam dilution effects are the same for both
H_2O and HDO, and that T_{KIN} is the same for both species.
Neither assumption can be made with complete confidence.
The ratio of the telescope diameter to wavelength is twice
as large for Waters et al.'s H_2O data as for Beckman
et al.'s HDO data. The latter authors give a diffraction-
limited beam size of ∿1.1 arcminutes, which is somewhat
larger than the CO core of the KL source, and so their
antenna temperature could be some 30% down on a beam-
corrected value, while for the H_2O very little beam
correction would be needed. At the same time, the identity
of T_{KIN} for the two species is unlikely, as from a
comparison of line widths the HDO emission seems to be
coming from nearer the core, i.e. from higher temperatures.
Selecting, however, a common value for T_{KIN} of 70 K one
obtains

$$|HDO|:|H_2O| = 1.35(\mp 0.5) \times 10^{-2} \qquad\qquad II$$

with the uncertainty coming mainly from the uncertainty in
the value of T_{KIN} itself.

It is in fact more realistic to split T_{KIN}, using
separate values for H_2O and HDO. For the HDO it is quite
difficult to assign a suitable value. Phillips et al.
(1977) using the $J = 3 \rightarrow 2$ transition of CO at 350 GHz give

T_{KIN} > 100K for the core, and Buhl et al. (1982) give
values greater than 100K from the J = 6→5 transition at
691 GHz.   Going to yet higher frequencies, Watson et al. (1980)
have measured lines of the CO J = 21→20 and J = 22→21
transitions at 2.4 THz and 2.5 THz respectively, which they
estimate to have been formed at T_{KIN} = 500-1000K, in the layers
just ahead of the shock-front emanating from the central
core (see Phillips and Beckman 1980).   This gives a range
of HDO formation temperatures of 1000K > T_{KIN} > 50K.   From
the velocity peak at 5.1 km s^{-1} we may narrow down this range,
and comparing this with the SO line observed at a similar
velocity, assign a value of T_{KIN} for the HDO 2(1.1) → 2(1,2)
transition of ∿100K.   We use the CO 'spike' temperature
of ≲50K, the temperature of the outer cooler cloud, for the
H_2O 3(1,3) → 2(2.0) transition.   In fact splitting the
temperatures in this way has a less striking effect on the
abundance than might be supposed.   Although in the case of HDO,
τ_{HDO} is halved, the ratio $(\frac{\tau}{N})_{HDO}$ is, from Fig. 1, reduced
by 1.5; hence N_{HDO} falls by only a factor 1.3 when T_{KIN}
is pushed from 50K to 100K.   The resulting abundance ratio is:

$$7.5 \times 10^{-3} < \{|HDO|:|H_2O|\}_{LTE} < 1.1 \times 10^{-2} \qquad III$$

with the range bounded by T_{KIN} values of 200K and 50K in the
core.   One must note that if $(T_{KIN})_{HDO}$ were placed at 1000K,
the resulting ratio would fall to 2.5 x 10^{-3}, thus encompassing
the lowest value, the $|DCO^+|:|HCO^+|$ value of Penzias.   If,
on the other hand, both lines were formed at 100K, our ratio
could rise to 1.8 x 10^{-2}.   Taking beam dilution explicitly
into account, and making the HDO come from a smaller
region than the CO core, and the H_2O from a larger region,
we can induce an upper value as high as 5 x 10^{-2} in the
ratio.   In this way we have obtained values for the
$|HDO|:|H_2O|$ abundance ratio ranging by as much as 20 in
value, depending on sets of physical condition which can
be quite plausibly assumed.
          In fact there are, even so, two serious omissions
from the present treatment: the presence of optical pumping

for either molecule, due to the proximity of the KL infrared source, and the use of a more comprehensive treatment of the radiative transfer within the source. Neglecting the former, there is some prospect of narrowing down the range by a careful application of the latter technique.

Conclusion

The ratio of $|HDO|:|H_2O|$ in the KL source lies between 5×10^{-2} and 2.5×10^{-3}, depending largely on the temperature distribution within a simple two-component model of the source. This range corresponds rather well with the range of values for abundance ratios between unsubstituted and singly deuterated molecules in the same source previously cited in the literature. It is therefore critically important to examine further the radiative and collisional formation of each line before making claims either about chemical differentiation onto grains, or theoretical constructs based on assumed real variations in the abundance ratios in different types of source.

References

Beckman, J. E., Watt, D. G., White, G. J., Phillips, J. P. and Frost, R. L., (1982),M.N.R.A.S. (In press).

Buhl, D., Chin, G., Koeff, G. A., Fetterman, H. R., Peck, D. D., Clifton, B. J., and Tannenwald, P. E.(1982) (This volume).

Penzias, A. A., (1979), Ap. J. 228, 430.

Phillips, J. P. and Beckman, J. E., (1980), M.N.R.A.S. 193, 245

Phillips, T. G., Huggins, P. J., Neugebauer, G., and Werner, M. W., (1977), Ap. J. 217, L161.

Rodriguez-Kuiper, E. N., Zuckerman, B., and Kuiper, T. B. H., (1978), Ap. J. 219, L49.

Snell, R. L., and Wootten, A. H.,(1979), Ap. J. 228, 748.

Turner, B. E., Zuckerman, B., Fourikis, N., Morris, M. and Palmer, P., 1975, Ap. J. 198, L125.

Wagoner, R. V., (1973), Ap. J. 179, 343.

Waters, J., Gustincic, J. J., Kakar, R. K., Kuiper, T. B. H., Roscoe, H. K., Swanson, P. N., Rodriguez-Kuiper, E. N., Kerr, A. R. and Thaddeus, P., (1980), Ap. J. 239, L125.

Watson, D. M., Storey, R. V., Townes, C. H., Haller, E. G., and Hansen, W. L., 1980, Ap. J. 239, L129.

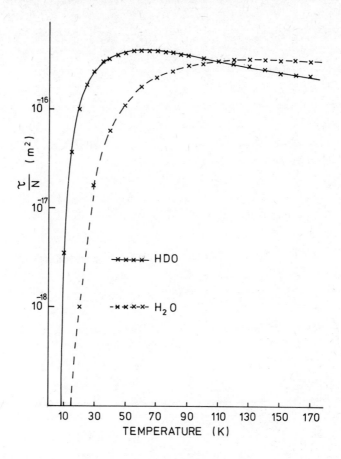

Figure 1 Ratios of optical depth to abundance for
the $2(1,1) \rightarrow 2(1,2)$ rotational emission line of HDO
and for the $3(1,3) \rightarrow 2(2,0)$ emission line of H_2O,
as in an LTE model.

DUST IN DENSE CLOUDS

ONE STAGE IN A CYCLE

J. Mayo Greenberg
Laboratory Astrophysics, University of Leiden, P.O. Box 9504
2300 RA Leiden, The Netherlands

Abstract
An evolutionary path is proposed for interstellar grains which takes them from diffuse clouds through molecular clouds and back as part of a repeating cycle in their development. Laboratory analog results are applied to the ultraviolet photoprocessing of grains in dense clouds. The cyclic occurrence of accretion, photoprocessing and destruction of grain mantles leads to the maintenance of a steady-state for the diffuse cloud grains in which the mantles consist primarily of a nonvolatile (refractory) complex molecular mixture containing oxygen, carbon and nitrogen but no H_2O. The organic refractory mantles shield the silicate cores against erosive processes while the production rate of silicate cores appears to be adequate against consumption in new stars. In the molecular cloud phase, explosions of grains containing ultraviolet produced frozen radicals balance against accretion to provide a quasi-steady-state between volatile molecules in mantles and molecules in the gas over a wide range of cloud densities. The mantles in molecular clouds are layered, the inner part being the organic refractory material. Models of layered dust grains are combined with laboratory measurements of the variability of the 3 μm H_2O absorption in mixtures to show how the observed ice band may provide a diagnostic probe of the evolution of grains. Some negative observational results are shown to be consistent with the layered model in which the outer layer contains a high percentage of H_2O. The strength and shape of the ice band in absorption in the protostellar source, B.N., are well matched by grains with rather thin outer mantles of ~ 0.03 μm as compared with the requirement of anomalously large grains based on pure amorphous ice mantles.

1. INTRODUCTION

Because the theme of this conference is submillimetre wave astronomy, the aspects of interstellar dust which are most relevant are those which prevail in molecular or dense clouds. Although it is the aim of this paper to provide just this information, the way to understand the nature of dust grains in dense clouds is to follow their evolution thorough all phases of the interstellar medium. Therefore I shall have

to start with grains in diffuse clouds. A number of basic properties of
the interstellar medium are summarized in Table 1. The chemical composition of the dust is constrained by the abundances of the elements.
The temperature of the dust is primarily determined by a steady state of
absorption and emission of the background radiation field so that in
most regions, exceptions as noted, the grain temperature is about 10 K
(Greenberg, 1971, Metzger et. al., 1982). This universal radiation field
produced by stars has a mean ultraviolet component in the tenuous
regions of space corresponding to a flux of $\Phi \approx 10^8$ photons cm^{-2} s^{-1}
(photons with energy $6 < E < 13.6$ eV) on any small body.

Table I. Average Interstellar Medium

Gas

$$0.1 \leftarrow \qquad \langle n_H \rangle = 1 \ cm^{-3} \quad \rightarrow 10^5$$

$$\langle n_{O+C+N} \rangle \approx 10^{-3} n_H$$

$$O : C : N \approx 6.8 : 3.7 : 1$$

$$\langle n_{Mg+Si+Fe} \rangle \approx 10^{-4} n_H$$

$$Mg : Si : Fe \approx 1 : 1 : 1$$

$$10,000 \ K \leftarrow T_{gas} = 100 \ K \quad \rightarrow 10K$$

Radiation in Ultraviolet

$$\langle n_{\lambda < \lambda_t} \rangle = 3 \times 10^{-3} \ cm^{-3}$$

$$\lambda_t = 2000 \ \text{Å} : h\nu_t = 6 \ eV$$

Dust

$$\langle n_d \rangle \approx 10^{-12} \ n_H$$

$$T_{dust} = 10 \ K$$

$$\bar{a}_d \approx 0.12 \ \mu m$$

$$[a_{core} = 0.05 \ \mu m, \ \bar{a}_{mantle} = 0.12 \ \mu m]$$

$$a_{bare} \approx 0.005 \ \mu m$$

$$n_{bare} \approx 10^3 \ n_d$$

The fact that oxygen, carbon and nitrogen are the most
abundant condensable species in space led Van de Hulst (1949) to derive
his "dirty ice" model for interstellar grains in which the saturated
molecules H_2O, CH_4, NH_3 would constitute the bulk of the grain, with
water dominating. We now have a variety of observational evidence for

the presence and absence of ice - as seen from its 3.08 μm absorption -
in a number of sources (see Fig. 13). Although it apparently exists in
dense regions where grains might be expected to be larger by accretion,
there does not appear to be any correlation between the strength of the
ice band and the size of the grains as determined from deviations from
normal extinction and polarization (Whittet and Blades, 1980; McMillan,
1978; Whittet, Bode, Evans and Butchart, 1981). Even over long distances
through diffuse clouds with total visual extinctions of greater than 10
magnitudes, I think we can say that the ice band "does not exist"
(Whittet, 1981). I shall show how these observations may be derived
naturally for a grain which evolves via accretion of oxygen, carbon and
nitrogen but which does not necessarily consist mostly of H_2O.

      Whether or not the expected H_2O absorption is found, there
is often a strong absorption at 9.7 μm corresponding to the Si-O stretch
in silicates, although the exact nature of the silicates is not fully
established. We shall not be concerning ourselves with this but merely
take as given that some sort of silicate particles are in space.

      In the framework of the grain model in which atoms and mole-
cules accrete on all congenial surfaces - some are not - the silicate
particles provide the nucleation cores for condensation from the gas.
The current estimates of the total mass loss rate from late type stars
is perhaps $dM/dt \sim 2.5 \times 10^{-12}\ M_0\ pc^{-3}\ yr^{-1}$. (Kwok, 1980). If this is so,
and all silicon by cosmic abundance is bound up in these M Supergiant
winds in the form of solid particles, the total mass supplied in 5×10^9
years is consistent with the mean lifetime of a grain (Greenberg, 1978).
This can not be much greater than 5×10^9 years - based on the current
star formation rate or equivalently the fact that 1/50 of the interstellar material is converted to stars each 10^8 yrs (Oort, 1974). Kwok
has reviewed the current status of theories of creation of refractory
grains and concludes that while M stars probably provide all or most of
the silicate material, other types of particles such as graphite,
amorphous carbon, silicon carbide are condensed in the winds of novae,
Wolf-Rayet stars and central stars of planetary nebulae. Our consideration is limited to the silicates which are the cores of the core-mantle
(core plus accreted O + C + N) particles.

      If we accept as inevitable that an atom (other than H or He)
or molecule which hits a 10 K grain (of sufficient size) sticks (Watson,
1976) then it becomes difficult to see why molecules exist in dense
clouds. It may be shown that the (e-folding) accretion lifetime for a

molecule of mass M = 30 (H_2CO) or M = 28 (CO) (at kinetic temperature T
= 50 K) is approximately (Greenberg, 1978)

$$\tau_{ac} \simeq \frac{3 \times 10^9}{n_{[H]}} \text{yrs}$$

where $n_{[H]}$ is the number density of hydrogen in all forms ($n_{[H]} = n_H +$
$2n_{H_2}$, for example). Therefore for $n_{[H]} = 10^4$ cm^{-3} the accretion time is
only ~ 3 x 10^5 yrs which is even shorter than the free fall time and is
certainly short compared to the minimum mean lifetime of 5 x 10^7 years
for a molecular cloud (Blitz and Shu, 1980). Another problem which bears
on this is how can one produce very complex molecules if depletion on
grains takes place. Let us consider for example the rate of production
of complex molecules by radiative association in the gas phase versus
the rate of sticking on grains. Starting with molecules M_1 and M_2
($M_2 > M_1$) we find that the ratio of the rates of combination of M_1 with M_2
and M_1 sticking to dust is

$$\frac{r_{M_1 M_2}}{r_{M_1 d}} \simeq \frac{n_{M_2} \; \sigma_{M_1 M_2}}{n_d \; \sigma_{M_1 d}}$$

where the association cross section $\sigma_{M_1 M_2}$ is $\simeq \pi a_{M_2}^2$ (Smith and Adams,
1977) and $\sigma_{M_1 d} \simeq \pi a_d^2$. Using reasonable values of a_{M_2} and a_d and assuming a concentration for the more massive molecule to be $(n_{M_2}/n_{[H]}) \simeq 10^{-9}$
(formaldehyde, for example) and the "standard" $n_d/n_{[H]} = 10^{-12}$ we get
the relative rate of recombination to sticking to be $r_{M_1 M_2}/r_{M_1 d} < 0.004$,
which is indeed small. How then, if grains accrete, do we get molecules
in the gas phase? I shall describe a mechanism by which grains may eject
accreted molecules back into the gas phase in molecular clouds.

The observational evidence for grain accretion is taken from
the variation of the wavelength dependence of extinction and polarization. For a summary of these data on variation see Greenberg (1978) and
Savage and Mathis (1979). The parameters commonly referred to are the
wavelength of maximum linear polarization λ_{max}, and the ratio of total
to selective extinction R = A(V)/E(B-V). In diffuse clouds the mean
value of λ_{max}^{-1} is λ_{max}^{-1} = 1.85 μm^{-1} and the mean value of R is R = 3.1. It
may be shown (Greenberg, 1976) that λ_{max} is related to the mean charac-
teristic dust radius by

$$\lambda_{max} \sim (m'-1) \; \bar{a} \qquad\qquad (1)$$

where m' is the real part of the refractive index. Thus λ_{max} increases with increasing particle size. From general scattering properties a relationship may be proven for dielectric particles such that λ_{max} and R must vary together with particle size (Greenberg, 1978). Empirically the relationship is R $\simeq$ 5.8 λ_{max} (Serkowski et al., 1975; Whittet and Van Breda, 1978). There are some outstanding exceptions to this correlation (Schultz et al., 1981) but, at this time, let us accept the concept of correlated extinction and polarization as stated above.

In the following sections the observational characteristics of grains will be derived from a model of interstellar grains in which the principal physical processes will be examined during the lifetime of a grain. Among the processes considered will be: effects of ultraviolet photons on and in the grains such as photoprocessing and photodesorption: grain-grain collisions; sputtering-erosion; evaporation; explosions. The effects of these processes are highly variable, depending not only on the grain composition but also on the local grain environment. Both of these will be pictured in terms of evolution in time and space wherein the grains pass through diffuse and molecular cloud phases consecutively and continually.

Although we think we have a good general understanding of the numerous physical processes which govern grain evolution, some of the quantitative aspects may not be certain enough to provide a unique solution to the problem. The methodology adopted here is to try to present a kind of global solution in which within the framework of current best estimates of growth and destruction processes one can develop a model which is consistent with a very wide range of observations. I shall show how the growth and destruction mechanisms can maintain a steady-state between dust in diffuse and dense clouds.

The basic new information on which the critical aspects of the overall scenario depends comes from results of experiments in Laboratory Astrophysics at the University of Leiden. A summary of these results is given in the next section.

2. SUMMARY OF LABORATORY PHOTOCHEMICAL RESULTS
(a) The astronomical problem – the laboratory solution

Grains interact with atoms, molecules and electromagnetic radiation in space. Let us consider first the physical and chemical interactions with electromagnetic radiation and, in particular, with electromagnetic radiation in the vacuum ultraviolet. Assuming, for the moment, that a grain of ~ 0.1 μm size consists of a dirty ice mixture of H_2O, CH_4, NH_3 what would happen to it in space? The ultraviolet photons with energies sufficient to photolyse these molecules would penetrate the particle (only ~ 1-2% of the photons are absorbed in the surface) and break the molecular bonds (Greenberg, 1978). This is schematically illustrated at the top of Fig. 1. Considering the reaction H_2O + hν → OH + H, we see that for photon energies greater than the photo-dissociation energy the hydrogen atom could come flying off and even

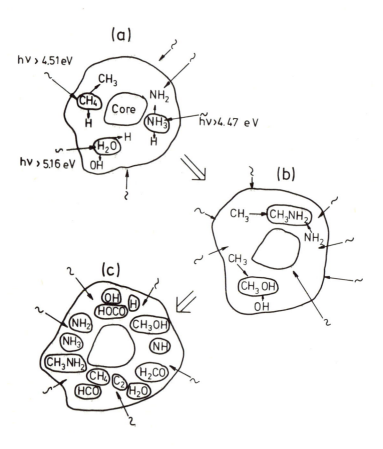

Fig. 1. Schematic evolution sequence for a grain mantle at 10 K subjected to ultraviolet photolysis. The processes illustrated are photodissociation, radical-radical combination, production of new molecules and radicals.

escape from the grain leaving the frozen radical OH – frozen because of
the low grain temperature. Similarly for the other molecules. In general
the energies required to photodissociate molecules are of the order of 4
eV or greater. Typical bond dissociation threshold energies for some
molecules and radicals are given in Table 2 (Calvert and Pitts, 1966).
The frozen radicals are chemically very reactive and should two of them
be adjacent to each other they would combine with zero activation energy
and release energy to the grain. The possibility of recombination is
shown in the second frame of Fig. 1 where, combining the hydroxyl
radical with the methyl radical, leads to the new, and more complex
molecule CH_3OH. The continuation of this type of process leads to a
grain with new molecules and frozen radicals as pictured in the last of
the sequence shown in Fig. 1. The conditions in interstellar space
should lead to just such a phenomenon being important almost everywhere
because the time scales for its occurrence are generally very short
compared with the interstellar time scales. An extreme case is given by

Table 2. Some approximate molecule and radical bond dissociation
energies adopted from Calvert and Pitts, 1966.

Bond Broken $R-R'$	ΔE (eV)	Bond Broken $R-R'$	ΔE (eV)
$OH - H$	5.14	$CH_2CH - H$	4.54
$O - H$	4.40	$CH_3 - CH_2O$	0.54
$O - O$	5.16	$OCH_2 - H$	1.02
$S - O$	5.44	$HCO - H$	3.79
$OS - O$	5.66	$H_2C - O$	7.59
$H - H$	4.50	$CH_3CO - H$	3.79
$C - O$	11.17	$H - CO$	0.755
$OC - O$	5.48	$C_2H_5 - H$	4.27
$CH_2 - CO$	2.32	$CH_3 - CH_3$	3.62
$CH_3 - OH$	3.84	$N - O$	6.52
$CH_3 - H$	4.40	$NH_2 - H$	4.47
$C - H$	3.53	$NH - H$	4.14
$CH - H$	5.48	$N - N$	9.78

placing the grains in the diffuse medium where the ultraviolet flux given in Table 1 for $E_{h\nu} > 6$ eV is $\Phi_{D.C.} = 10^8$ cm^{-2} s^{-1}. I use a lower limit of 6 eV as a convenient and perhaps conservative critical value for a photodissociation energy. A lower limit of 4.5 eV results in a doubling of the ultraviolet flux (Greenberg, 1978). The time required for each bond in a grain to have seen a photon with $E > 6$ eV is given by the number of bonds in the grain divided by the total number of photons absorbed per unit time. This photoprocessing time (τ_{pp}) is approximately given by

$$\tau_{pp} = \frac{4/3 \; \pi \; a^3 e^{-\tau_{uv}}}{d^3 \Phi_{DC} \pi a^2} = \frac{4 a e^{-\tau_{uv}}}{3 d^3 \Phi}$$

where d is a bond diameter and τ_{uv} is an ultraviolet attenuation factor. For a = 0.1 μm , d = 3 Å the value of τ_{pp} in the diffuse cloud medium ($\tau_{uv}=0$) is only about 200 years which is indeed very short. The production of complex molecules and radicals in grains leads to physical and chemical processes which play a role not only in determining the chemical constituents of the grains but also the chemical constituents of the gas. In order to understand and quantify the relevant phenomena one must study the effect of ultraviolet photons on materials and under conditions which prevail in interstellar space. Early attempts to do this at temperatures (T ≃ 28 K) approaching that of the grains and with photon energies ~ 7.5 eV (Greenberg et al., 1972, Greenberg, 1973) proved that this method would work. Similar experiments were performed at rather high temperatures (T = 77 K) and somewhat lower photon energies (E < 5 eV) but their consequences were also interesting (Khare and Sagan, 1973). One of the principal aims of the laboratory developed at Leiden is to simulate as closely as possible the important interstellar conditions in order to study the chemical and physical evolution of dust.

The Astrophysical Laboratory at the University of Leiden which was established in 1975 is the first to succeed in simulating the essential interstellar space conditions as they affect the evolution of interstellar grains. A schematic of the main elements of the experimental set-up is shown in Fig. 2. The key components are the low temperature and the ultraviolet. The low temperature is achieved by means of a closed cycle helium cryostat within which one reaches temperatures as low as 10 K on a "cold finger" which can variously be an aluminium block or transparent window mounted on a metal ring. Various gases may be controllably allowed to enter the vacuum chamber of the cryostat (pressures

down to 10^{-8} torr) via a capillary tube. These gases condense as a solid
on the cold finger which acts then like the core of the interstellar
grains. On one port to the chamber is mounted a source of vacuum ultraviolet radiation which until now has almost exclusively been a microwave
stimulated hydrogen flow lamp which has emission peaks at 1216 Å (Lyα)
and about 1600 Å. The normal flux of vacuum UV photons by these lamps
is ~ 10^{15} cm^{-2} s^{-1} at the target. Through another port (or pair of
ports) we may direct the beam of an infrared spectrometer which

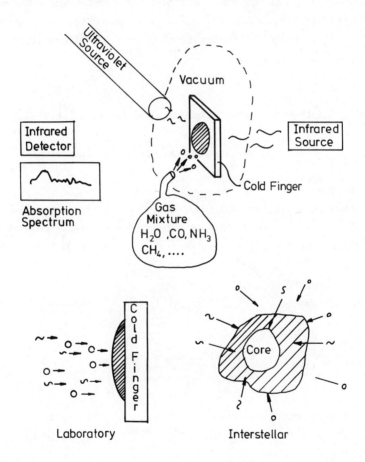

Fig. 2. Schematic of the laboratory analog method for studying
interstellar grain evolution. Molecules are deposited as a
solid on a cold finger in a vacuum chamber and irradiated by
ultraviolet photons. The infrared absorption spectrum shows the
appearance and disappearance of various molecules and radicals.
The cold finger may be an aluminium block (~ 3 cm) or a glass,
sapphire or LiF window.

measures the infrared absorptions in the sample on the cold finger
between 2.5 μm and 25 μm (4000 cm^{-1} to 400 cm^{-1}). This is the "finger-
print" region for identifying molecules by their stretching, bending and
rocking modes of ocsillation in a solid. Other monitoring measurements
are of pressure, chemiluminescence, mass spectra and visible absorption.
Further details of the equipment may be found elsewhere (Hagen,
Allamandola and Greenberg, 1979).

A comparison between laboratory and interstellar conditions
is seen in Table 3. The most important - but necessary - difference is
in time scales for photolysis. Relative to accretion rates of molecules
the flux rate in the laboratory is such that in diffuse clouds one hour
is equivalent to 1000 years.

Table 3. Comparison between laboratory and interstellar conditions

	Laboratory	Interstellar
Grain mantle		
- initial composition	CO, H_2O, NH_3, CH_4..	All condensible species
- thickness	0.5 μm to 10 μm	≈ 0.1 μm
- temperature	⩾ 10 K	⩾ 10 K
Gas: pressure of condensible species	10^{-7} mbar	$3n_{[H]} \times 10^{-20}$ mbar
Ultraviolet flux $\lambda < 2000$ Å	10^{15} cm^{-2} s^{-1}	10^{8} cm^{-2} s^{-1}
Time scales		
- Diffuse clouds	1 hr.	10^{3} yrs.
- Molecular clouds	1 hr.	~ 10^{4} - 10^{6} yrs.

The basic mode of operation consists of deposition of
mixtures of simple volatile molecules —CH_4, CO, H_2O, CO_2, NH_3, N_2, O_2—
and simultaneous irradiation as they freeze on the cold finger. Sometimes irradiation is continued after deposition is stopped. We simulate
in this way the accretion and photoprocessing of grains in molecular
clouds. The principal laboratory sequences and operations are:

1. Infrared absorbtion spectra of irradiated pure samples and mixtures at 10 K. Infrared studies of warmed up (annealed) and recooled samples.

2. Infrared spectra following irradiation to detect new molecules and radicals.

3. Infrared spectra of irradiated material following warm up to follow disappearance of frozen radicals and formation of new molecules.

4. Visible and ultraviolet absorption spectra of irradiated and warmed up samples.

5. Simultaneous measures of chemiluminescence (visible) and vapor pressure during warm up of irradiated and, for comparison, unirradiated samples.

6. Production of explosions in warm up period.

7. Infrared and mass spectrometric analyses of complex non-volatile residues remaining after warm up to room temperature.

8. Visible absorption spectra of non-volatile residues.

I shall indicate briefly in the following some sample results from the laboratory. More extensive treatments will be presented elsewhere.

(b) In Situ Production Of New Molecules And Radicals – Infrared Spectra

In Fig. 3 are shown some examples of results of the laboratory analog sequences of dust evolution. The absorption spectra in Fig. 3 show first an unirradiated sample and then, for comparison, the spectra after irradiation showing the appearance of new molecules and frozen radicals. We see, for example, that molecules like formaldehyde and formamide are readily created and it may be inferred that much more complicated molecules are also being produced at the low temperatures. Their presence is clearly indicated after warm-up as shown in the upper sequence where, as the more volatile molecules are evaporated, the absorption spectrum takes on the very different character shown in the two upper right spectra. Both samples show that the HCO radical is easily created and probably plays an important role in subsequent stages. The sample illustrated in the lower half shows that H_2CO is produced in the cold solid in quantities which may even become comparable (in this sample) with the H_2O and NH_3 as indicated by the relative absorption intensities in the 1600 cm^{-1} region.

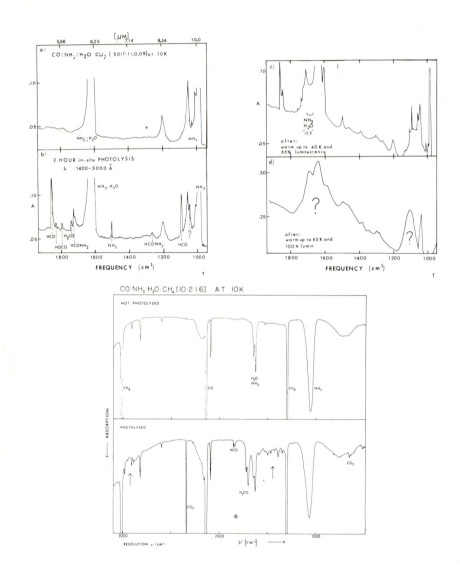

Fig. 3. Infrared absorption spectra of two sample analog grain mantles. Left side of upper sequence and the two lower spectra show first the features in unirradiated samples and then the spectra of the irradiated samples showing the appearance of new molecules and radicals produced by photoprocessing. Upper right hand pair of spectra clearly indicate the presence of complex molecules (unidentified) which appear as the more volatile species are evaporated away by warming up.

(c) Explosions of Irradiated Samples

It was early demonstrated from both chemiluminescence and pressure enhancement in warmed up irradiated materials that energy was released not only in visible but in heat form by radical-radical or radical-molecule interactions. Only about 10^{-5} of the energy is released as visible light. The fact that these reactions are diffusion controlled is also clear from the fact that the luminescence stops immediately upon cool down and does not resume until the warmup again reaches the temperature at which the luminescence is stopped. Explosive events can be systematically produced in the laboratory (d'Hendecourt, et al., 1982) by ensuring that the reaction energy is not conducted away from the sample too rapidly. An example of such an explosion is shown in Fig. 4 illustrated by simultaneous pressure spikes along with light flashes. When such events occur, essentially all the sample is blown off of the cold finger. It has been noted that the explosions appear to occur for a variety of different samples at temperatures of T = 27 K. From infrared measurements of the shape of the NH_3 absorption in the insulating layer between the exploded material and the cold finger we have established that the temperature overshoots to at least ~ 70 K during the explosion thus clearly demonstrating independently the tremendous energy release which occurs. An important criterion for the explosion to take place is that the rate of ultraviolet photons hitting the sample be at least 1/10 of the rate of accretion of the condensing molecules (($dn_{h\nu}/dt$) / (dn_{mol}/dt) $\geq$ 0.1). This may be shown to imply a photolysis efficiency of about 10%.

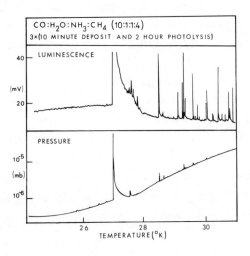

Fig. 4. Correlation of chemiluminescence spikes with pressure bursts from the mixture $CO:H_2O:NH_3:CH_4$ (10:1:1 :4) after three cycles of 10 minute deposition plus irradiation followed by 2 hours of additional photolysis at 10 K and subsequent warm-up.

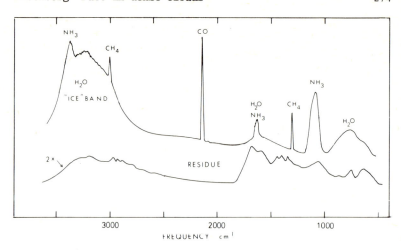

Fig. 5. Comparison of infrared absorption spectra of "yellow stuff"
residue with 10 K mixture containing the same amount of oxygen,
carbon and nitrogen in molecular form as in the initial (pre-
irradiated) residue material. Note the complete absence of an
H_2O ice band at 3.08 μm in the residue spectrum.

(d) Complex Organic Residue

As was illustrated in Fig. 1 and Fig. 3 during the photoprocessing of the grain mantle analog material more and more complex
molecules are created. When the volatile components in the sample are
evaporated away by warming there always remains a nonvolatile residue
material. If one starts with a cosmic abundance mixture of CO : H_2O :
NH_3 : CH_4 : the ultimate residue appears yellow. We have obtained
infrared absorption spectra of nonvolatile residues (hereafter referred
to as OR for organic refractory) with various initial compositions.
However we have not followed up on the photoprocessing by examining the
results of continued ultraviolet irradiation of the residues themselves.
Nevertheless what we have already done provides some important quantitative answers to the grain composition. One of our samples had a
molecular weight of 514 and all of our samples do not evaporate at
temperatures less than ~ 400 K with at least one pyrolyzing, without
evaporating, at 600 K. The infrared absorption spectrum of a residue is
shown in Fig. 5 and comparison is made with the spectrum of the original
unirradiated mixture. The relative absorption strengths have been normalized by equating the integrated absorptions of the unirradiated sample
and the residue between 2000 cm^{-1} and 1000 cm^{-1}. This region is chosen
to avoid the H_2O absorption enhancement which appears in the 3000 cm^{-1}

region. We identify the very broad absorption from 3500 cm^{-1} to 2000 cm^{-1} as due to carboxylic acid groups and a number of absorptions between about 1200 cm^{-1} and 1500 cm^{-1} as due to amino groups. See Fig. 6 for an example of the spectrum of a molecule identified as 8-carbomoyl adenine. The extra features around 3.4 μm are attributed to the C-H stretch in various combinations.

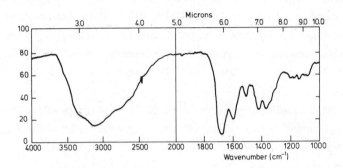

Fig. 6.   Infrared absorption spectrum of an HCN-oligomerization product after Sephadex G-15 fractionation and HPLC separation. Identified as 8-carbamoyl adenine (Figure courtesy of A.W. Schwartz and A.B. Voet).

A high resolution mass spectrometer analysis of the lowest pressure component in the residue whose infrared spectrum is given in Fig. 5 showed, after warmup and CO_2 release, a mass corresponding to $C_4H_6N_2$ and traces of urea. The intensity ratios suggest that amino pyroline rings make up a substantial part of this material. One possible representation as part of a polymer is:

Undoubtedly this material will undergo further modification when subjected to continued ultraviolet bombardment. However a very exciting property of the yellow stuff which gives direct evidence for something like it being a ubiquitous component in interstellar space is its visible absorption spectrum which exhibits an excellent likeness in both position and width to several of the unidentified diffuse interstellar bands (Merrill, 1934; Herbig, 1975) and particularly the famous λ 4430 band (Baas, as reported in Greenberg, et al., 1980). It is hypothesized that, under interstellar conditions of ultraviolet irradiation, a yellow stuff which contains traces of the metallic elements Fe, Mg, Mn, Ca will provide a major source of the some 50 observed interstellar visible absorption bands.

High molecular weight molecules in liquid or solid form have significantly higher real values of the index of refraction than the volatile ices. We have not yet measured this but a survey of the data for other complex molecules shows that we can expect a value of m' of at least m' = 1.40 and perhaps as high as m' ⩾ 1.5 (Handbook of Chemistry and Physics, 1966–67). I have adopted a preliminary estimate of m'_{OR} = 1.45. An additional new optical property of the OR material relative to the classical ices is that it begins to absorb in the near ultraviolet with significant absorption probable at wavelengths as long as λ ≃ 2500 Å (Calvert and Pitts, 1966; Silverstein and Bassler, 1967). This tends to suppress the extinction oscillations which would be exhibited by ∼ 0.15 μm size simple ices in the ultraviolet up to $\lambda^{-1} \simeq 5$ μm^{-1} (see, for example the calculated curves in Fig. 9).

(e) The H_2O Ice Band

Interpretation of the observations of the 3 μm ice band (O–H stretch) requires a knowledge of the absorptive properties of H_2O in various mixtures and at various temperatures relevant to interstellar dust. The first complete measurements of pure solid H_2O (Bertie et al., 1969) provided an important guide to the early observations. However because they were made for pure crystalline ice rather than for ice as it occurs naturally in interstellar space they led to some apparent inconsistencies in shape and position of the ice band which even led to suggestions that the ice band may not be due to H_2O at all (Mukai et al., 1978).

In the Leiden Astrophysics Laboratory it has been possible to study ices under conditions which match those of interstellar space.

We have started first with pure H_2O ice even though generally H_2O must occur in mixtures along with other molecules in interstellar grains. This work has served as a bench mark or standard with which to compare various mixtures prepared under similar conditions. These studies have appeared in detail in a number of publications (Hagen, Allamandola and Greenberg, 1980; Hagen, Tielens and Greenberg, 1981; Hagen, Tielens and Greenberg, 1982). I shall summarize here a few of the critical results with emphasis on the spectral features around 3 μm.

One of the most important aspects of interstellar ice is that it forms and exists mostly at extremely low temperatures, T = 10 K. The H_2O ice deposited very slowly at this temperature is extremely amorphous (there are various degrees of amorphicity) and accounts for the fact that the O–H absorption is about twice as broad as that of crystalline ice (Fig. 7). Annealing the sample up to 80 K (still amorphous) results in both a shift in peak absorption and an increase in

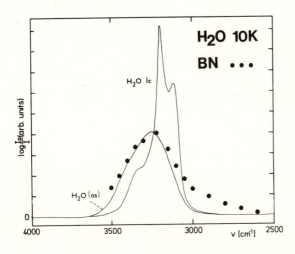

Fig. 7.
Absorption of amorphous ice H_2O(as) at 10 K and crystalline ice H_2O (Ic). The shape of the ice absorption in BN is shown by the dots.

intensity (at the peak) as well as a decrease in width. Thus ice properties measured at liquid nitrogen temperatures (Léger et al., 1979) differ noticeably from those at 10 K. It is important to note that the librational absorption at ~ 800 cm^{-1} of unannealed H_2O is shifted to lower frequencies, broadened and reduced in intensity relative to that of annealed forms of solid H_2O and that these effects are enhanced in mixtures. This is of relevance to the fact that the "normal" (or crystalline) H_2O absorption at 12 μm exhibits strong changes in the 9.7 μm silicate band (Greenberg, 1978) which have not been detected.

Thus two of the apparent inconsistencies between observations and measurements of the O-H absorption - the width and position
- are accounted for by amorphous H_2O ice. The long wavelength wing which
is seen in almost all sources (an exception being OH 231.8 + 4.2) is
reproduced by depositing other molecules along with H_2O in various
mixtures. Two examples of measurements of optical contants of mixtures
which include NH_3 are shown in Fig. 8. It should be pointed out that
ammonia is not the only molecule which can produce the wing, although
there is good reason to expect its presence from the double peaked shape
of some of the observed 3 µm bands (see Fig. 13, section 5 and Willner
et al., 1982). In Table 4 a summary of the optical properties of H_2O in
various mixtures deposited and measured at 10 K is presented.

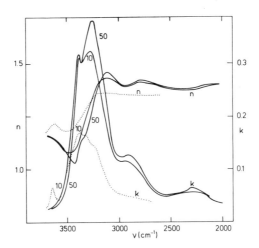

Fig. 8.   Real and imaginary parts of the index of refraction
for two ice mixtures. Solid curves for $H_2O:NH_3$ = 3:1, Dotted
curves are $H_2O:NH_3:CO$ = 5:1:6. Note that n-ik here is the same
as m'-im" in the present paper. These data are taken from
Hagen, Tielens and Greenberg, 1982.

Table 4. Optical constants of various ice mixtures at peak absorption [a] (~ 3 μm).

f_{H_2O} [b]	Form	Half width	m'	m"	$m_f''/m_1'' \equiv f_{\text{eff}}$
1	Crystalline	150	1.37	0.815	
1	amorphous (10K)	310	1.31	0.477	1
0.75 [c]	amorphous (10K)	320	1.29	0.324	0.68
0.42 [d]	amorphous (10K)	350	1.25	0.170	0.36
0.20 [e]	"	$\sim$350	$\sim$1.25	0.024	0.05
0.15 [e]				0	0

(a) Peak values of m". Values of m" at peak absorptivities for small particles differ slightly.

(b) $f_{H_2O} \equiv$ fraction of H_2O in mixture

(c) $H_2O : NH_3 = 3 : 1$

(d) $H_2O : NH_3 : CO = 5 : 1 : 6$

(e) Extrapolated by a mean straight line through measured points.

3. GRAINS IN DIFFUSE CLOUDS

I shall be assuming that there are essentially two size populations of interstellar grains – a bimodal model. The justification for this is illustrated in Fig. 9. The smoothed average interstellar extinction curve and partial curve for the average interstellar polarization are compared with the polarization and extinction of perfectly aligned infinite dielectric cylinders with index of refraction m = 1.3 – 0.05 i. The scales of the two curves have been adjusted so that the value of λ^{-1} (in μm^{-1}) in the observed curves corresponds to the value of $2\pi a/\lambda$ in the calculated curves. We may immediately deduce a typical particle size which reproduces the maximum polarization by inserting $(\lambda^{-1}_{max})_{obs} \simeq 2$ in $(2\pi a/\lambda_{max})_{calc} \simeq 2$. This gives $\bar{a} \simeq (1/2\pi)$μm $\simeq$ 0.15 μm for those particles which give the polarization and extinction in the visual. This is not too different from a size derived from more detailed models. Can these also give the far ultraviolet extinction? The answer is no; the reason being again a general particle scattering property. We note that the extinction by the cylinders saturates at some

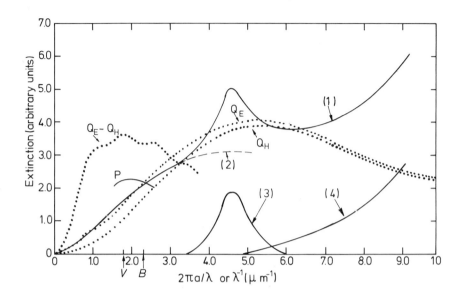

Fig. 9. Comparison of theoretical extinction curves for aligned circular cylinders with index of refraction m=1.3-0.05i with
observed average interstellar extinction (curve 1) and polarization (p). Curves 2 and 4 are separated contributions to the
total extinction by classical sized (a ≃ 0.15 μm) and small
(a ≃ 0.01 μm) particles. Curve 3 is the "2200 Å" bump. Q_E and Q_H
are the extinction efficiencies (per unit length) for cylinders
lined up along and perpendicular to the electric vector of
incident linear polarized radiation. Theoretical extinction is
$(Q_E + Q_H)/2$ and polarization is $Q_E - Q_H$.

value of $x = 2\pi a/\lambda \simeq 5$ and, except for some broad oscillations, does not
again rise. In fact, since the value of $x > 4$ corresponds to ultraviolet wavelengths for a = 0.15 μm and since all materials become absorbing in the ultraviolet the extinction will actually be reduced from
that which is shown. A number of examples may be seen in Greenberg
(1978). Thus the far ultraviolet extinction can not be produced by the
same grains which provide the visual extinction. It is for this reason
that one must divide the extinction curve for the visual and ultraviolet
into two or three distinct parts as shown in Fig. 9 where the grains
which provide the visual extinction and polarization are labelled "2"
and those which provide the far ultraviolet extinction are labelled "4".
Those which provide the "2200 Å" bump may be entirely different from
those which give the far ultraviolet extinction and are therefore

labelled separately as "3". We may estimate the particle size for the far ultraviolet in a similar way to what we used for the visual region. Although I haven't shown it here, the characteristic extinction curve for absorbing particles (see for example Greenberg, 1978) is such that in the region $2\pi a/\lambda \simeq 0.5$ the curve is rising as is the observed one at $\lambda^{-1} \simeq 9$ μm^{-1} We derive the characteristic particle size to be $a_b \simeq 0.01$ μm where the subscript "b" stands for bare and implies that these particles can not accrete mantles of atoms and molecules from the gas (Greenberg and Hong, 1974; Purcell, 1976; Greenberg, 1978). We shall assume, for the moment, that the particles which produce the 2160 Å bump are some form of carbon or organic material and are different from the particles which produce the far ultraviolet extinction. These latter are taken to be silicates. The bare particles will not be considered much from an evolutionary point of view. We shall however carry them along for completeness where needed in such questions as cosmic abundance and depletion and ultraviolet attenuation in clouds.

It is to the grains which produce the visual extinction that I shall now focus attention. We start with the classical idea that these grains consist of silicate cores with accreted mantles of the O, C, N group (organics) in some combination with hydrogen. A model based on this assumption has been developed rather fully in terms of core-mantle cylinders (Hong and Greenberg, 1980). Since all attempts at observing the H_2O ice band absorption in diffuse clouds have proven negative, it is assumed that H_2O ice is not an abundant constituent in the diffuse cloud grain mantles even though oxygen is the most abundant of the organic condensable. Since the volatile organics have been shown in the laboratory to evolve into a refractory component in times which are astronomically short we shall suppose, for the moment, that the mantles consist of this material. A fuller account of grain accretion supporting this hypothesis is developed in the next section. We shall here examine the consequences of this hypothesis in terms of the H_2O absorption.

The absorption strength of an organic refractory mixture at $\sim$ 3 μm can be estimated from Fig. 5 to be about 1/6 that of a cosmic abundance mixture containing H_2O; namely $m''_{O.R.} \simeq 0.19/6 \simeq 0.03$. The low value of this absorption in combination with the much broader shape (800 cm^{-1} as compared with $\sim$ 300 cm^{-1} for H_2O) make this absorption in the 3 µm region difficult to observe. It has been shown (Greenberg, 1981) that using a core radius of a_c = 0.05 µm and a mantle radius of a_m = 0.12 µm, the ratio of 3 µ absorption to visual extinction is

$$\frac{A^{OR}_{obs}(3)}{A(V)} \simeq 0.01$$

This means that for an extinction of 20 magnitudes the expected absorption by the OR mantle is only 0.20 and very broad, therefore difficult to detect.

The narrower structure in the OR mantle should be detectable. A number of infrared sources in the galactic center have been used to probe the spectrum of the intervening material and the evidence is clear for absorption features in the 2.8 µ to 4 µ region some of which undoubtedly are produced in diffuse clouds between us and the galactic center (see for example Wickramasinghe and Allen, 1980; Willner et al., 1979). For the very broad feature referred to the evidence is not clear and work is in progress on comparing the observed features with spectra of various laboratory residues (Van de Bult, Allamandola, Greenberg, 1982). What is clear is that some sort of complex molecular mantles exist on the grains and we are studying in the laboratory how these complex materials further evolve in time and how they are affected by the constant immersion in the relatively high flux ultraviolet field which prevails in the diffuse clouds.

We conclude, at least, that the negative evidence for H_2O in diffuse clouds can readily be ascribed to the existence of complex organic mantles which can contain some oxygen but which have a very different absorption spectrum from H_2O.

In Table 5 I have summarized the distribution of the O, C, N and Si, Mg, Fe (refractory) elements as they are observed or inferred in the various solid and gaseous components in diffuse clouds. It is assumed that no molecules are present in the gas and that O, C, N, Si, Mg, Fe gaseous abundances are as observed in ζ Ophiucus (Morton, 1974; Jenkins and Shaya, 1979; de Boer, 1980). The abundances in the core and bare particles are estimated from Hong and Greenberg (1980). For the mantle we let the outer radius a_m = 0.12 µm which is at the lower limit of our previous model results because the OR index of refraction in the visible is substantially larger than that of ordinary ices. Since the mean particle size is governed by the dimensionless parameter $a(m'-1)$ = const., an increase in m' implies a decrease in a (Greenberg, 1970,

1978). Instead of cosmic abundance proportions of O, C, N in the mantle we use the values implied by the mass spectrum results for the organic refractory of Fig. 4 (O:C ≈ 1:2). This is a preliminary estimate and may later be modified with further experimental results but it does not appear likely that much less oxygen or much more carbon are possible except in almost pure hydrocarbons. Note that the total amount of oxygen

Table 5. Elemental composition of diffuse clouds relative to cosmic abundances.

	O	C	N	Mg	Si	Fe	
Gas							
Atoms + Ions	0.75	0.20	0.20	0.03	0.03	0.01	(a)
Molecules	--	--	--	--	--	--	(b)
Dust							
Core + bare	0.09	0.27		~ 1.0	~ 1.0	~ 1.0	(c)
Mantle							
OR	0.11	0.42	0.22				(d)
Volatiles	~ 0	~ 0	~ 0				
(H_2O, etc.)							
Total Gas	0.75	0.20	0.20				
Total Solid	0.20	0.69	0.22				
Unaccounted	0.05	0.11	0.58				
Available for							
accretion (Av.)	0.80	0.31	0.78				

a) Depletions in ζ Oph from Morton (1974), Jenkins and Shaya (1979) and from de Boer (1980).
b) Negligible
c) The carbon depletion is based on a 0.025 μm radius graphite particle to produce the 2200 A bump. Should another type, or smaller size of particle be the actual cause, the carbon depletion in the bare particles will be substantially reduced from the value 0.27.
d) Based on the assumption of a mantle material with relative atomic composition as given by the mass spectrum results for the organic refractory in section 2d.

depleted in the grains is probably less than ~ 0.2 and that the carbon depletion is accordingly ~ 0.7. This means that when grains grow in molecular clouds they must grow largely as a result of the oxygen accretion and we may therefore anticipate water rich mantles appearing. It is seen that a substantial fraction of carbon is needed to produce the 2200 Å bump. Should this feature be produced by some other candidate material it is probable that the available gas phase carbon will have to be revised upwards.

4. EVOLUTION OF DUST GRAINS

(a) General Scheme: The Molecular Cloud-Diffuse Cloud Cycle

The aim of this section is to integrate the laboratory investigations into a scenario which follows the chemical evolution of dust grains. Further elaborations and justifications are taken up in later sections.

A schematic of several stages in the evolution of a dust grain is shown in Fig. 10. We start with the birth of a grain assumed here to be in the form of an elongated silicate particle of ~ 0.05 μm radius. These small particles are swept up into the gaseous matter in space after being blown out of cool stellar atmospheres and then begin to partake in the evolution of clouds as they pass through the various conditions and environments of which we observe a snapshot at any one time. From the fact that the clouds can not be static — either because they are observed to be in motion, or because we infer or see such dramatic energetic events as star formation occurring within them — it is obvious that the physical conditions of density and temperature represent different stages in their evolution (see middle column of Fig. 10). The densest clouds seem to correspond to times just before, during or just after star formation. Diffuse clouds become dense by a number of mechanisms, perhaps as a result of collisional combinations or some source of external pressure (Field and Saslaw, 1965; Kwan, 1979; Oort, 1954; Scoville and Hersch, 1979; Taff and Savedoff, 1972a; 1972b). Within the dense clouds critical densities may be reached which lead to instabilities and further contraction and finally to star formation (Woodward, 1978; Bash, 1979). After the stars form — if they happen to be large hot stars or if they develop high material ejection speeds by processes other than radiation (accretion disks for example) — the remaining local material from which they have been formed is ejected

into the surrounding space (Blitz and Shu, 1980; Lada and Harvey, 1981).
Much of this material, being heated and finding itself in a very tenuous
low pressure environment expands to reappear as diffuse clouds.

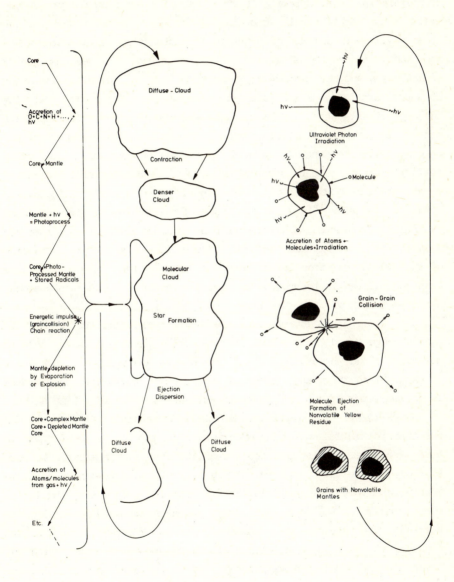

Fig.10.   Diagram of grain evolution. The sequence on the left
         corresponds to one of the subcycles in the dense cloud phase.
         The sequence on the right is a contraction of how grains evolve
         through the molecular cloud and star formation phase and back
         to the diffuse cloud phase.

Should the silicate particles appear first in a diffuse cloud region it is certain that no gaseous material will accrete (or remain) on them principally because of photodesorption by interstellar ultraviolet photons. Sticking of atoms or molecules from the gas on the grains only begins within molecular clouds where it proceeds simultaneously with the process of ultraviolet photolysis. We are now at the start shown at the top of the left sequence in Fig. 8. After some period of accretion and photoprocessing two grains will collide with each other at suprathermal speeds sufficient to make the grains explode. Following the explosions – which may be complete or partial – the grains now have mantles of various composition and thickness all the way down to the original silicate core. In general, a residue of complex organic material will be built up on the silicate cores, each successive generation within the cloud leading to an additional layer. Since the indication from the laboratory is that the order of 2% to 20% of the condensed material is converted to the nonvolatile residue each 10^7 years we may assume that a substantial mantle thickness, say a significant fraction of the 0.07 μm required in the mean for diffuse cloud grains, will have been accumulated in the course of time the grain is in the dense cloud, which we assume to be of the order of 10^8 years.

We shall assume that the total cycle time for a grain to pass through the diffuse cloud (or intercloud) phase and the subsequent molecular cloud phase is ~ 2 x 10^8 years with about half of this spent in each. Since the mean lifetime of a grain (the refractory components) is of the order of 5 x 10^9 years we estimate that in this time a total of 10–20 cycles are repeated. We shall show that it then becomes possible to maintain a steady-state distribution of particle types in diffuse clouds. Furthermore it is statistically more realistic to have started (in the left sequence of Fig. 10) with a core-mantle particle than with a bare core particle. In the next few sections we shall look more closely at the individual steps we have proposed above for the chemical and physical processing of grains. Whatever processes are expected to take place in each region are limited to those which can occur in the maximum time interval of ~ 10^8 years unless otherwise stated.

(b) Grains in Molecular Clouds

In molecular clouds, grains experience both growth and destruction. The growth process carries with it its own destruction mechanism which generally prevents the grains from exhausting the condensable species from the gas. Let us first consider growth.

It has been proposed that grain explosions could resupply accreted molecules to the gas (Greenberg, 1973a,b; Greenberg and Yencha, 1973; Greenberg, 1976). Subsequently the suggestion was made that the triggering mechanism for the grain explosions was most likely to be the collisions between grains at suprathermal speeds produced by the turbulent gas motion (Greenberg, 1979). Both of these suggestions have been followed up using the laboratory results on the conditions for explosive reactions (d'Hendecourt et al., 1982). The basic laboratory derived parameters are: a critical grain temperature T_c = 27 K; a critical ultraviolet photon absorption rate relative to molecular collision rate of $r_c \geqslant 0.1$. Using a specific heat of 0.1 joule g^{-1} (a reasonable specific heat in the 10 - 30 K range) gives a critical grain-grain collision speed of v_c = 40 m s^{-1} to raise the temperature from 10 to 27 K. Since the time required for a grain to radiate its heat away at this temperature is $\sim$ 10 sec there is adequate diffusion of frozen radicals to complete the chain reactions. The turbulent gas speeds of $\geqslant$ 1km s^{-1} in molecular clouds are supersonic (Larson, 1981) but it is not entirely clear what grain-grain collision speeds these imply. If one applies the formulation of Völk et al. (1980) which is correct only for subsonic gas speeds, then a turbulent speed of 1 km s^{-1} gives a grain-grain collision speed (for equal sized grains $\sim$ 0.1 μm in radius) of 100 m s^{-1} which is well above the critical speed. We use this result keeping in mind that a proper theory of grain-grain collisions in a supersonic turbulent medium remains to be developed. If we balance the accretion rate on grains with the molecule ejection rate by explosions (of $\sim$ 100% efficiency) we may show that the fraction, f, of available condensable volatile atoms (or molecules) which remain in the gas is (d'Hendecourt et al., 1982) f = 40-60% where the available atoms of O, C and N are those remaining after the dust depletion in diffuse clouds are subtracted from cosmic abundances (Table 5). We are assuming that the diffuse cloud phase passes continuously into the molecular cloud phase for both dust and gas.

The grain explosion process will be operative as long as

$$r = \frac{\Phi_{D.C.} \; e^{-\tau_{uv}}}{n_{mol} \langle v_{mol} \rangle} > r_c = 0.1 \qquad (2)$$

where $e^{-\tau_{uv}}$ is an effective attenuation factor for the ultraviolet flux inside the molecular cloud. Another possible limiting factor to the occurrence of explosions is that the grain-grain collision rate may be too rapid to allow an adequate thickness of irradiated material – a critical thickness (Greenberg, 1976) – to accumulate between collisions. If we assume no internal sources of ultraviolet radiation such as provided by shocks (Schull and McKee, 1979) or by stellar winds (Silk and Norman, 1980) then τ_{uv} is a mean optical depth for the photons. Approximating τ_{uv} by $\tau_{uv} \simeq 2A(V)$ (Greenberg, 1979) and letting $\langle v_{mol} \rangle \simeq 10^4$ cm s^{-1}, the above condition gives the results shown in Table 6 where I have used a diffuse cloud flux value for photons with energy greater than 6 eV of $\phi_{D.C.} = 10^8$ cm^{-2} s^{-1} (Greenberg, 1978). This is some 30% less than the value suggested by Metzger et al. (1982) but the conclusions should not be substantially modified. It appears that the accretion/explosion balance applies almost everywhere and that only when $n_H > 10^5$ cm^{-3} will accretion dominate. On the other hand, if as estimated by Metzger (1982, private communication), the flux within 3kpc from the galactic center is

Table 6. Relative collision rates of ultraviolet photons and molecules in homogeneous diffuse clouds (D.C.) and molecular clouds (M.C.). Standard parameters: $\phi_{h\nu}^{(a)} = 10^8 e^{-2A(V)}$ cm^{-2} s^{-1}, $n_{mol} = 10^{-3}$ $n_{[H]}$, $v_{mol}(D.C.) = 4 \times 10^4$ cm s^{-1}, $v_{mol}(M.C.) = 1.5 \times 10^4$ cm s^{-1}, $Y_{p.d.} = 2 \times 10^{-5}$ ($\sigma_{p.d.} \simeq 10^{-20}$ cm^2).

A(V) for Explosions ($n_c \gtrsim 0.1$)	$n_{[H]}$ (cm^{-3})	A(V) for Accretion ($n_{p.d.} < 1.0$)
7.4	10	0.80 [b)
6.6	50	0
5.8	10^2	—
4.6	10^3	—
3.4	10^4	—
2.3	10^5	—

a)   Flux reduced by additional factor 1/2 in clouds with $n_{[H]} \gtrsim 10^2$ cm^{-3} to account for reduced solid angle (see text).

b)   A spherical cloud 10 pc in radius with $n_{[H]} = 10$ cm^3 has an extinction at the center of $0\overset{m}{.}16$ so that A(V) = 0.8 is unlikely to occur.

an order of magnitude or more greater, the grains and gas have a significantly different relative distribution towards the galactic center.

A rather interesting corollary of the accretion/explosion phase is that the very small bare particles of the bimodal distribution may act very much like large molecules in relation to the larger core-mantle grains. By this I mean that, just as the molecules collide and stick to the grains, the bare particles may also collide and stick and become imbedded in the volatile mantle material. When the grain mantle explodes, the bare particles return to the gas along with the ejected molecules. Since the turbulence of the medium implies a grain-grain collision speed $\sim 10^4$ cm s^{-1} which is comparable with the thermal molecule-grain collision speed, the rate of accretion of bare particles would be similar to that of the molecules and therefore, just as a fraction f $\sim$ 50% of the volatile molecules is bound to the grains, we expect a substantial fraction of bare particles also bound. One obvious implication of such a phenomenon is that the ultraviolet attenuation within the cloud would be substantially reduced. This is doubly effective because eliminating a portion of the very small particles gives rise to an increased albedo and a higher value of the scattering asymmetry factor g = $\langle$cos $\theta \rangle$ both of which permit deeper penetration of the ultraviolet into the cloud (Sandell and Mattila, 1975; Flannery et al., 1980). This effect should be considered in a more complete treatment and it would be useful to observe whether the far ultraviolet extinction by the small particles in molecular clouds is smaller than in the diffuse medium.

In addition to the "steady-state" mantle of volatiles produced by photoprocessing we expect a gradual increase in the amount of nonvolatile organics. Some of this material will be ejected into space -perhaps as small chunks $\sim$ 0.01 µm in size-but some should remain. The best estimate we can make at this time is that, with a conversion rate of 2-20% per 10^7 years, between $\sim$ 20% and 100% (!) of the processed material may be added to the grains each time they pass through the molecular cloud phase. Since observations require a steady-state between molecular cloud and diffuse cloud phases this will have to be balanced by some erosion of the OR material while the grains are in the diffuse cloud phase. This will be taken up later.

Inside molecular clouds, the grains are generally shielded from the most effective destruction mechanisms. The only two which appear to be important for grain mantle materials are photodesorption and sublimation by supernova explosions (Draine and Salpeter, 1979a, 1979b, henceforth referred to as D-Sa,b) and then only for very loosely bound molecules. As has been discussed there at some length the photo-desorption efficiency is actually unknown to orders of magnitude and the best available estimate should not be taken too seriously. Based on L.T. Greenberg's (1973) experimental results, Greenberg (1974) assumed a value of the photodesorption efficiency for icy mantles of $Y_{pd} \simeq 10^{-5}$. This was substantially lower (by orders of magnitude than that taken by Barlow (1978) and by Watson and Salpeter (1972) but is about the same as can be derived from the photodesorption cross section $\sigma_{pd} = 10^{-20}$ cm^{-2} (for the 6 eV to 13.6 eV interstellar photons) now suggested (D-S, 1979). The photodesorption yield efficiency is defined as $Y_{pd} = \sigma/d^2$ where d^2 = area of the molecule. The condition for accretion to dominate over photodesorption in clouds is

$$r_{pd} = \frac{Y_{pd} \, \Phi_{D.C.} \, e^{-\tau_{uv}}}{n_{mol} \, \langle v_{mol} \rangle} \geqslant 1 \qquad (3)$$

where the photodesorption efficiency has been assumed to be $Y_{pd} = 2 \times 10^{-5}$. The results of this prediction are contained in Table 6 where it is seen that for $n_H \geqslant 50 \mathrm{cm}^{-3}$, and therefore essentially over the whole of a molecular cloud, accretion should be important and provide the basis for the accretion/explosion phenomenon via photoprocessing. Note that when the solution of equation 3 gives $A(V) \geqslant 1$ it is necessary to lower the effective flux by an additional factor (of about 2) to take account of the fact that the radiation is penetrating from only one side of the cloud rather than from all directions ($\Omega < 4\pi$). Neglecting accretion and considering photodesorption alone, volatile molecular mantles have lifetimes of $\sim 2 \times 10^8$ years, which is longer than the molecular cloud lifetime, and can certainly be ignored.

        The only remaining effective destructive mechanism is super-nova sublimation which, when it occurs, also ejects the dust outward from the molecular cloud. Considering its effectiveness only at such times, those grains which remain in the molecular cloud are not affected but those which are ejected will have every bit of their loosely bound mantle material (with binding energies $U_0 \leqslant 0.2$ eV) removed in the process. However, according to D-Sb materials like H_2O with $U_0 > 0.5$ eV are hardly affected and therefore we shall assume that they survive in the return passage to the diffuse cloud medium.

(c) Grains in Diffuse Clouds

The core-mantle grains which are ejected into the diffuse or intercloud medium have, in addition to an OR mantle, a possible additional volatile mantle of which (see section 5) H_2O is probably a dominant constituent. Even should a portion of more fragile material survive, it erodes away too rapidly to be significant. Materials like CH_4 will be evaporated away as a result of heating by supernova explosions at a volume erosion rate of $\tau_e^{-1} = 2 \times 10^{-5}$ yr^{-1} (D-Sb) which is equivalent to a radius erosion rate of

$$da/dt = (a_0/3\tau_e)e^{-t/3\tau_e} \qquad (4)$$

where a_0 is the radius at time $t = 0$.

This means that a mantle of thickness 0.03 μm lasts less than 10^5 yrs on a 0.15 μm radius grain. Both the initial grain size and mantle thickness were chosen as representative for observational reasons (see section 5). The supernova sublimation of H_2O is negligible and D-S conclude that in diffuse clouds erosion results mainly from high velocity cloud-cloud collisions, cloud "crushing" from supernove remants, and photodesorption. It is not clear that chemically bound mantle material can be photodesorbed in the same way as physically adsorbed atoms or molecules. Ignoring it leads to $\tau_e^{-1} \simeq 6 \times 10^{-8}$ yr^{-1} (by interpolation from D-S), and including it leads to $\tau_e^{-1} = 3 \times 10^{-8}$ yr^{-1} for the H_2O component. Therefore a thickness of 0.03 μm may last 1-2 x 10^7 yrs. At first sight this seems to imply that there is some possibility of observing H_2O in diffuse cloud grain mantles. However, if I include the effects of intense diffuse cloud photoprocessing (which changes the H_2O into other less detectable forms) and the H_2O detection criteria shown in Table 8 (next section) the observability is severely reduced.

In the above I have not considered the intercloud medium where destruction rates are far faster so that if a significant fraction of grains (and gas) pass through this phase, the mean time scales for survival may be shorter than estimated. On the other hand if, as it seems likely, no more than a few percent of all the interstellar matter is in the intercloud phase only a small fraction of the grains will be affected. In terms of the standard cloud model with a density $n_H \simeq 20$ cm^{-3} and a filling factor $\simeq 0.1$, (Spitzer, 1978) and including the fact that perhaps 50% of the total interstellar material is in molecular clouds with densities $n_H > 10^3$ cm^{-3} (Gordon and Burton, 1976) it is

difficult to conceive of much material left over after considering the
mean value of $N_H/E(B-V) = 6 \times 10^{21}$ cm^{-2} mag^{-1} (Spitzer, 1978).

Once the more volatile mantle components have been cleared
away the underlying mantle of OR material is exposed to the same des-
tructive processes. Very little is known about the sputtering yields for
complex molecular substances. A semi-empirical formula for low energy
sputtering yields was developed by Draine in 1977 for normal incidence
($\theta=0$) (which we adopt for application to the OR mantle) is

where
$$Y = A \frac{(\epsilon - \epsilon_0)^2}{1 + (\epsilon/30)^{4/3}} \qquad \epsilon > \epsilon_0$$

$$\epsilon \equiv \eta E U_0^{-1}, \; \eta \equiv 4 \, \xi \, M_P M_T (M_P + M_T)^{-2}$$

$$\epsilon_0 \equiv \max \, [1, 4\eta]$$

where M_P = projectile mass, M_T = target mass, U_0 = binding energy of
molecule (or atom), $\xi \approx 1$, $A \approx 8.3 \times 10^{-4}$.

Assuming a moderately large mean molecular weight of
$M_T = 300$ for the OR material and a binding energy $U_0 > 2eV$, the yield
for $E = 300$ eV hydrogen projectiles is $Y_{OR} \lessapprox 2 \times 10^{-4}$. This is
comparable with the yield for silicates with $U_0 = 5.7$ eV which is
$Y_{SIL} \approx 10^{-4}$. Thus, approximating the OR volume erosion rate as $\tau_e^{-1} \approx$
10^{-8} yr^{-1} we find that the radius of the organic refractory may be
decreased by as much as 30% in 10^8 years. This appears to be sufficiently small to be replenishable (see section 4b) in each molecular
cloud phase so that it is reasonable to assume a mean OR mantle in the
diffuse clouds.

The spectra of galactic center sources of which those shown
in Fig. 11 are a recent sample, exhibit no evidence of H_2O ice although
very broad absorptions peaked at around 3 μm are not uncharacteristic.
Is this related to the broad OR absorption or is it perhaps due to a
further photo-evolved material? The features around 3.4   μm are
undoubtedly produced by C-H stretches in various molecules and configurations. These, and other grain mantle absorptions are being studied in
the Leiden Astrophysics Laboratory for the purpose of relating them to
laboratory produced samples of photochemical evolution of which a
preliminary example is shown in Fig. 5.

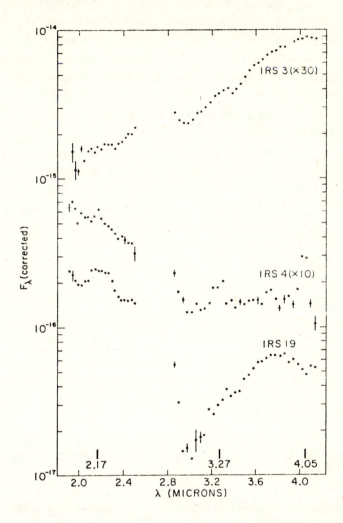

Fig. 11. Spectra of galactic center sources corrected for interstellar extinction (taken from Willner and Pipher, 1982). Note the variety of absorptions peaked at about 3 µm but evidently not due to H_2O.

5. CHEMICAL COMPOSITION OF DUST AND GAS IN MOLECULAR CLOUDS

In this section I shall consider the composition of grains in molecular clouds to have evolved by accretion of mantles on grains which originated in diffuse clouds. Our base, therefore, consists of an organic refractory mantle on a silicate core. It is upon this already established mantle that an <u>additional</u> mantle of volatile constituents is accreted. I shall use the H_2O absorption band as an observational probe of the volatile mantle.

Although H_2O ice is seen in the grain spectra in molecular clouds it has not been possible in the past to measure its abundance because of a lack of accurate data on the complex indices of refraction of ice mixtures in the 3 μm region. Now, we may use the laboratory data as given in section 2.

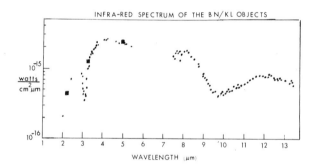

Fig. 12. An early infrared absorption spectra of the BN source.

I have chosen as my key example the well observed H_2O absorption (and polarization) in the B.N. object. The 3.08 μm optical depth deduced from Fig. 12 is A(3.08) = 1.46. The total extinction has been derived to consist of two parts: about 30 magnitudes or less due to the cool dust in the outer cocoon and about 27 magnitudes due to the interior dust heated to about 400 K (Bedijn, 1977). The hot dust must have evaporated off all its volatile constituents but it can retain the organic refractory mantle. It is however possible for an extra mantle of volatiles to be found on the cool dust. For this component we then have the observed ratio of ice absorption to visual extinction

$$\left[\frac{A(3.08)}{A(V)} \right]_{obs} = \frac{1.46}{30} = 0.05 \qquad (5)$$

The question is, what fraction of the outer mantle is H_2O and how thick is this mantle. It is known that the dust in Orion exhibits extinction and polarization wavelength dependences characteristic of larger than average size grains. We can deduce that an extra mantle thickness of ~ 0.03 μm on the diffuse cloud grain will produce the shift in the λ_{max} of the polarization to correspond to that in Orion (Breger, 1974).

The absorption for a homogeneous spherical (the spherical shape is chosen for calculational convenience) grain of radius a is given, for $2\pi a/\lambda < 1$, by (see for example Greenberg, 1978)

$$A_{abs}(\lambda) = 9k \; 4/3 \; \pi a^3 \; \frac{\epsilon_2(\lambda)}{(\epsilon_1+2)^2 + \epsilon_2^2} \qquad (6)$$

where $\epsilon_1 = m'^2 - m''^2$, $\epsilon_2 = 2m'm''$, $(m = m' - im'')$, $k = 2\pi/\lambda$.

Since we are here dealing with core-mantle particles, the absorption by the mantle should be calculated from a rather more complicated formula (Van de Hulst, 1957). However, exact calculations for core-mantle particles (Greenberg and Hong, 1974) empirically justify the following simple extension of the Rayleigh approximation to a layered mantle particle with the extra absorptivity in the outer layer:

$$A_{abs}(\lambda) = 9k \; 4/3 \; \pi (a_2^3 - a_1^3) \; \frac{\epsilon_2(\lambda)}{(\epsilon_1+2)^2 + \epsilon_2^2} \qquad (7)$$

where a_1 is the radius of the inner OR mantle and a_2 is the radius of the outer mantle. The small fractional volume of the core with $a_c = 0.05$ μm is neglected, being 10% of the inner mantle volume.

The extinction per grain in the visual, V, is

$$A(V) = Q(V) \; \pi a_2^2 \qquad (8)$$

where for normal, or near normal, grain sizes, $Q(V) = 1.5$.

The ratio of absorption to extinction (ignoring the small differences between optical depths and magnitudes) is

$$\frac{A_{abs}(\lambda)}{A(V)} = 8k \; \frac{a_2^3 - a_1^2}{a_2^2} \; \frac{\epsilon_2}{(\epsilon_1+2)^2 + \epsilon_2^2} \qquad (9)$$

Let us consider, for trial purposes, that 75% of the mantle is H_2O. Then, with $m''_{0.75}$ = 0.32, m' = 1.29 (Table 3), we get from equation 9,

$$[\frac{A_{abs}(3.08)}{A(V)}]_{0.75} = 0.07 . \tag{10}$$

It thus appears that between 50% and 60% (0.05:0.07 = 0.54) of the mantle is H_2O ice! At first sight this may appear to be almost inconsistent with the fact that, by cosmic abundance, the fraction of O relative to O + C + N is only 0.58 because it implies that all of the oxygen in the mantle is bound up in H_2O. However, as we noted in the depletion table for diffuse cloud grains (Table 4) the available fraction of oxygen in the volatiles which enter the molecular cloud phase is $\simeq$ 0.72 which is substantially larger than the cosmic abundance value, 0.58. Therefore the result above says that a large but reasonable fraction of the available oxygen is in the form of H_2O.

We define F_2 as the fraction of available O+C+N in the volume of the extra mantle contained within a_2 and a_1 denoted by $(O + C + N)_{Av}$. Since, in the model above, this volume is very close to that of the base mantle $(a_2^3 - a_1^3 \simeq a_1^3 - a_c^3)$ we may write

$$F_2(O+C+N)_{Av.} \simeq F_1(O+C+N)_{C.A.} \tag{11}$$

where F_1 = 0.22 (see Table 4, under "available for accretion" and $(O + C + N)_{C.A.}$ = cosmic abundance of O + C + N. This gives F_2 = 1.56 F_1 = 0.34 so that about 35% of the available volatiles are in the extra grain mantle. This is consistent with the prediction in section 4 based on the steady-state accretion/explosion process (d'Hendecourt et al., 1982). Actually this comparison may not be completely relevant because the dust in the foreground of B.N. is more likely to be post- than pre-star formation dust. However another, and more direct consistency check of mantle size has been made by calculating the shape of both the 3.08 μm absorption and polarization and comparing with observations (Hagen, Tielens and Greenberg, 1982) (see Fig. 14.). It has been shown there that although the long wavelength wing in the absorption in the BN source in Orion may be reproduced by pure amorphous H_2O ice grains of abnormally large size , (~ 0.4 μm is required to provide the additional scattering) such grains do not give a match to the polarization. However grains whose size is a = 0.15 μm which contain H_2O diluted with NH_3

(other strong bases can also produce a low frequency wing) reproduce the shapes of the polarization as well as the extinction.

In Table 7 I have summarized the expected abundances in the various components of a molecular cloud based on the above model for the dust. The range of observable molecular densities with respect to hydrogen seems to be quite small (with CO as the notable exception) when compared with the elemental abundances. For example, the total carbon in all carbon bearing molecules each with i carbon atoms and molecular species M_j is

$$\frac{[n_C]_{mol}}{[n_H]} = \sum_{M_j^i} i \, [M_j^i]/[n_H] \tag{12}$$

If we use a representative value of $i = 3$ in the ~ 50 molecules with $0 < i < 9$ observed (Mann and Williams, 1980) and a representative abundance $[\bar{M}]/[n_{[H]}] = 10^{-8}$ based on H_2CO (probably an overestimate), we deduce that the total observed relative abundance of carbon in molecules, still excluding CO is

$$\frac{[n_C]_{mol}/[n_{[H]}]}{[n_C]_{C.A.}/[n_{[H]}]} < \frac{50 \times 3 \times 10^{-8}}{3.7 \times 10^{-4}} = 0.004 \tag{13}$$

which is much less than the ~ 0.10 found in CO (Wootten et al., 1978).

Some examples of how the entries in Table 7 are arrived at are as follows:

$$\text{O in solid } H_2O \quad = \frac{0.8 \times 0.34 \times 5.41}{6.76} = 0.22$$

$$\text{O in other solids} \quad = \frac{0.2 \times 0.34 \times 5.41}{6.76} = 0.05$$

$$\text{C in other solids} \quad = \frac{0.34 \times 1.15}{3.7} = 0.11$$

One notable feature in Table 7 is that it predicts a substantial amount of oxygen and nitrogen in hitherto undetected molecules. An obvious candidate for the nitrogen is N_2 which is quite stable. A candidate to take up a part of the oxygen is CO_2. While the O_2 molecule might be suggested it is perhaps less likely because it photodissociates at rather lower energies. All of these molecules are not observable by radio detection because they have no dipole moments.

Table 7. Elemental composition of B.N. type molecular clouds relative to cosmic abundances.

	O	C	N	Si	Mg	Fe	
Gas							
Atoms + Ions	—	—	—	—	—	—	(a)
CO	0.05	0.10	—	—	—	—	
	(~0.01)	(~0.03)					(b)
Other Molecules	<0.01	<0.01	<0.01				(c)
Dust							
Core + bare	0.09	0.27		~ 1.0	~ 1.0	~ 1.0	
Mantle							(d)
Solid H_2O	0.22						
OR	0.11	0.42	0.22				
Other	0.05	0.11	0.26				
Total Gas	0.06	0.11	0.01				
Total Solid	0.47	0.80	0.48				
Unaccounted	0.47	0.09	0.51				
Available for accretion	0.53	0.20	0.52				

a) Not counting possible significant carbon ions as in Phillips et al. (1980).

b) If depletion of gas CO in dense cores of molecular clouds (Rowan-Robinson, 1979).

c) Observed (see text).

d) Assumes extra grain mantle of 0.03 μm and the ice absorption for B.N. (see text).

        While ice mantles are predicted by the above theory for molecular clouds this does not necessarily imply that they are easily observable. A case in point is the star HD 29647, a reddened early-type star in the Taurus dark cloud. According to Whittet et al. (1981) there is about 3 magnitudes of extinction for this star and the ratio of total to selective extinction is $R = \dfrac{A(V)}{E(B-V)} = 3.5 \pm 0.1$ as compared with the interstellar mean $R = 3.1$. This means the grains are larger than average and should therefore have an extra mantle possibly containing a substantial fraction of H_2O. We estimate the higher outer grain radius from $\Delta R/R_{IM} \sim \dfrac{\Delta a}{a_{IM}}$ which gives a grain size about 10 percent larger than average. Using an H_2O fraction in this extra mantle ($a_2 - a_1 = 0.1a_1$) of 0.72 (the maximum possible) we get

$$\left[\frac{A_{abs}(3.08)}{A(V)} \right]_{0.72} \approx 0.03 \tag{14}$$

from which we find

$$\frac{E(3.08 - V)}{E(B-V)} \approx -3.40$$

which (see Whittet et al., 1981, where this quantity is used as the comparison parameter) is consistent with the apparently negative observation of the 3.08 µm absorption. It should be noted that Whittet et al. (1981) expect a much larger H_2O absorption because they have used a grain model different from mine in two important aspects: (1) They have assumed a mantle consisting entirely of H_2O (no OR submantle), (2) Their ice absorptivity per unit mantle mass is about two times too large because it does not take into account both the proper degree of amorphicity as well as the dilution of the H_2O in the mantle mixture.

We shall now derive a simple general expression for the ratio $A(3)/A(V)$ as a function of the value of $R = A(V)/E(B-V)$ and f_{H_2O}, the degree of H_2O dilution. We approximate the variations in R as proportional to grain size so that $R'/R \approx a_2/a_1$, noting as before that a_1 is the radius of the inner mantle which contains no H_2O. Let $R'/R = X$ then Equation 9 may be rewritten as

$$\left[\frac{A(3)}{A(V)} \right]_{X,\, f_{H_2O}} = 0.18 \frac{X^3-1}{X^2} f_{eff} \tag{15}$$

where f_{eff} is the effective fraction of H_2O producing absorption at 3 µm and f_{eff} is $\sim m''$. I am making use of the empirically derived fact that to within $\sim 5\%$ the peak ice absorptivity of small spherical grains is directly proportional to the value of the imaginary part of the index of refraction. We see in Table 4 that the measured values of m'' (and of

f_{eff}) extrapolate to 0 at f ≈ 0.15. This is because the O-H stretching modes in monomeric and dimeric bonded aggregates occur at around 2.7 μm (~ 3700 cm^{-1}) rather than at around 3.1 μm and means that even a 15% concentration of H_2O is too dilute a mixture to allow enough clumping of H_2O polymerization to produce the 3 μm solid H_2O (as) absorption. This is understandable because for, say f = 0.10, there are enough solvent molecules to completely surround (in terms of nearest neighbours) each water molecule, thus statistically isolating them from each other. For example, in a simple cubic lattice the concentrations of singles, doubles and triples at a concentration p of the solute molecule are (Behringer, 1958)

$$n_S = p\,(1-p)^6$$
$$n_D = 3p^2\,(1-p)^{10} \qquad\qquad (16)$$
$$n_T = 3p^3\,(1-p)^{13}\,[4+(1-p)].$$

For p = 0.15 only about 10% of all the solute molecules form triples so that the expected value of the relative absorptivity of H_2O (at 3.1 μm) is $m_T''/m_1'' \leqslant (0.1)(0.15) = 0.015$. This qualitatively confirms the estimates in Table 4 for the absorptions at small concentrations.

Table 8. Ratio of ice absorption to visual extinction, A(3)/A(V), for various mantle thicknesses and H_2O concentrations.

X \ f_{H_2O}	0.15	0.20	0.30	0.40	0.50	0.60	0.70
1	0	0	0	0	0	0	0
1.1	0	0.005	0.008	0.014	0.019	0.025	0.032
1.2	0	0.010	0.016	0.027	0.037	0.048	0.060
1.25	0	0.012	0.019	0.032	0.044	0.057	0.071
1.3	0	0.014	0.022	0.037	0.051	0.066	0.083
1.4	0	0.018	0.027	0.046	0.064	0.083	0.104
1.5	0	0.021	0.032	0.050	0.076	0.099	0.124

This effect has been observed in the laboratory (Hagen, Tielens and Greenberg, 1982). The variations of the predicted ice absorption to visual extinction are presented in Table 8 for several values of X and f to show a range of reasonable possibilities for different kinds of dust clouds. The case for B.N. is represented by X = 1.25 and f ≈ .55. The negative example for the star HD 29647 is represented by X = 1.1, for which the maximum H_2O content yields the results obtained earlier. It is interesting that the positive ice absorption $A(3)/A(V) \approx 0.04$ for RCrA (Whittet and Blades, 1980) requires both a substantial fraction of H_2O and a substantially larger than average R - perhaps like the values for B.N. and for NGC 2024 No. 2 (Merrill et al., 1976). The principal evidence for high H_2O content in grain mantles seems to be limited to protostellar sources (see Fig. 13) of which BN has been the classical example (For some additional examples see Willner et al., 1982). These represent a state of the grains just after star formation as modified from the state just before. Both of these phases are probably different from that which is required to provide the basis for the quasi steady-state equilibrium between accretion and explosion of grains. In the latter, if the grain mantles have acquired, by accretion or surface reactions, a substantial concentration of H_2O it should appear as abundantly as CO in the gas phase and this is not observed (Waters et al., 1980). Thus, within the framework of the accretion/explosion phenomenon (in at least some molecular clouds), the mantles contain relatively little H_2O - certainly much less than CO. This is apparently inconsistent with the first model calculations of the molecular composition of grain mantles (Tielens and Hagen, 1982) which take into account both gas phase and surface reactions but which do not yet include photoprocessing effects. Further work is being done to include all three phenomena.

Such negative results as are obtained in the Ophiucus dark cloud are apparently explainable in terms of rather small mantle concentration of H_2O. All null results (Whittet and Blades, 1980) like that for HD147889 can, of course still imply as much as 15% H_2O, the question now to be answered being on what does this fraction depend. Both the manner of accretion and the photolysis of the mantle material - the past history of the local grains - can probably provide sufficient variability. Actually I make no claim to understand the star HD 147889 because its extinction characteristics are so peculiar everywhere - in the visible, the near ultraviolet and the far ultraviolet. It is possible that local phenomena in the very dense small dust cloud surrounding this star, including coagulation of dust, may play a role in producing the apparently anomalous properties.

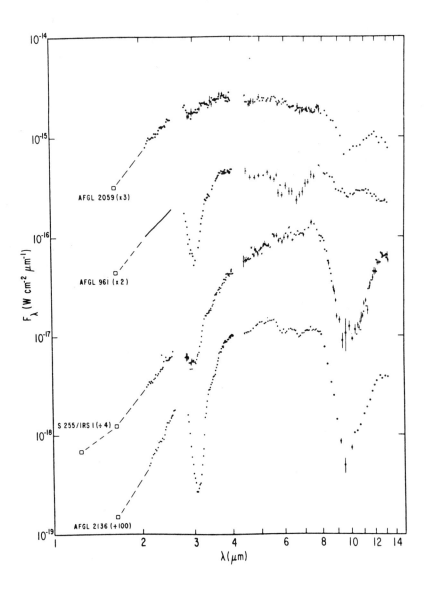

Fig. 13. Infrared absorption spectra of several protostars (from Willner et al., 1982).

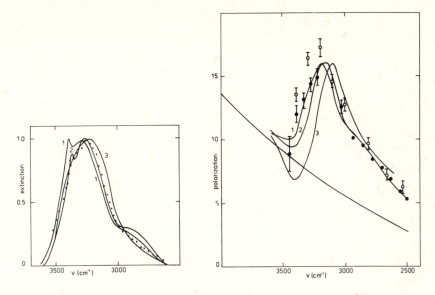

Fig. 14. Comparison of the calculated cross sections (3250 cm^{-1} feature) of infinite cylinders with: (a) the absorption and (b) the polarization towards BN.
1.   Unannealed mixture of H_2O/NH_3 (3/1) (T = 10 K). Mean size 0.15 μm.
2.   Partially annealed mixture of H_2O/NH_3 (3/1) (deposited at 10 K warmup to 50 K and recooled to 10 K). Mean size 0.15 μm.
3.   Unannealed H_2O(as) (10 K). Mean size 0.4 μm.
Dots (Gillett et al, 1975); open circles (Kobayashi et al., 1980); filled circles (Capps et al., 1978).

In Figure 14 a comparison is made between observations and model calculations for grains which contain an H_2O mantle component.

The conclusion we come to is that although the looked for correlation between the strength of the H_2O ice band and the growth of mantles is not a simple one we can begin to conceive that its variability can provide a deeper insight into processes in clouds other than simple accretion.

Acknowledgements

I gratefully acknowledge the comments and criticisms of all the members of the Laboratory Astrophysics group who patiently listened to several seminars based on various aspects of the work summarized here. Their stamp appears indelibly on the contents, although they should not be considered responsible for the final product.

REFERENCES

Bash, F.N., 1979, Astrophys. J., 233, 524-538.
Bash, F., Hausman, M. & Papaloizou, J., 1981, Astrophys. J., 245, 92-98.
Barlow, M.J., 1978, M.N.R.A.S., 183, 397.
Bedijn, P.J., 1977, Ph.D. Thesis, Leiden University.
Behringer, R.E., 1958, J. Chem. Phys., 29, 537.
Bertie, J.E., Labbé, H.E. & Whalley, E., 1969, J. Chem. Phys., 50, 4501-4520.
Blitz, L. & Shu, F.H., 1980, Astrophys. J., 238, 148-157.
De Boer, K.S., 1980, Astrophys. J., 224, 848.
Breger, M.. 1974, in Planets, Stars and nebulae Studies with Photo-polarimetry ed, T. Gehrels, Univ. of Arizona Press, Tucson, 946.
Van de Bult, C.E.P.M., Allamandola, L.J., & Greenberg, J.M., 1982, paper in progress.
Calvert, J.G. & Pitts, J.N. Jr., 1966, Photochemistry, J. Wiley & Sons, N.Y.
Capps, R.W., Gillett, F.C. & Knacke, R.F., 1978, Astrophys.J., 226, 863.
Draine, B.T. & Salpeter, E.E., 1979a, Astrophys. J., 231, 77.
Draine, B.T. & Salpeter, E.E., 1979b, Astrophys. J., 231, 438.
Field, G.B. & Saslaw, W.C., 1965, Astrophys. J., 142, 568-583.
Flannery, B.P., Roberge, W. & Rybicki, G.B.,1980, Astrophys.J.,236, 598.
Gillett, F.C., Jones, T.W., Merrill, K.M. & Stein, W.A., 1975, Astron. Astrophys., 45, 77.
Gordon, M.A. & Burton, W.B., 1976, Astrophys. J., 208, 346-353.
Greenberg, J.M., 1970, in Interstellar Gas Dynamics, ed. H. Habing, D. Reidel, Dordrecht, 306.
Greenberg, J.M., 1971, Astron. & Astrophys., 12 240.
Greenberg, J.M., Yencha, A.J. Corbett, J.W. & Frisch, H.L., 1972, Ultraviolet Effects on the Chemical Composition & Optical Properties of Interstellar Grains, Mem. Soc. Roy. Sciences Leige, 6e serie, Tome III, 425-436
Greenberg, J.M., 1973a, Symposium on the Origin of the Solar System, Ed. H. Reeves, Edition du Centre Nat. de la Rech. Scientifique, 135-141.
Greenberg, J.M, 1973b, Molecules in the Galactic Environment, eds. M.A. Gordon & L.E. Snyder, J. Wiley, N.Y., 94-124.
Greenberg, J.M. & Yencha, A.J., 1973 in Interstellar Dust and Related Topics, Ed. J.M. Greenberg, & H.C. van de Hulst, D. Reidel, Dordrecht, 369-373
Greenberg, J.M., 1974, Astrophys. J. (Lett.),. 189, L81-L85.
Greenberg, J.M. & Hong, S.S., 1974, in HII Regions and the Galactic Center, Ed. A.F.M. Moorwood, ESRO, SP-105, 153-161.
Greenberg, J.M., 1976, Astrophys. and Sp. Sci., 39, 9-18.
Greenberg. J.M., 1978, Cosmic Dust, Chapter 4, ed. J.A.M. McDonnell, J. Wiley & Sons Ltd., 187-294.
Greenberg, J.M., 1978b, In Infrared Astronomy, ed., G.S. Setti & G. Fazio, D. Reidel, Dordrecht, 51-97.
Greenberg, J.M., 1979, in Stars and Star Systems, Ed. B.E. Westerlund, D. Reidel, Dordrecht, 173-193.
Greenberg, J.M., Allamandola, L.J., Hagen, W., Van de Bult, C.E.P.M. & Baas, F., 1980, Interstellar Molecules, IAU Symposium no. 87, ed. B.H. Andrew, D. Reidel, Dordrecht, 355-363.
Greenberg, J.M., 1981. Ned. Tijd. voor Natuurkunde, A 47 (1): 24-26.
Hagen, W., Allamandola, L.J. & Greenberg, J.M., 1979. Astrophys. & Sp. Sci., 65, 215-240.

Hagen, W., Allamandola, L.J. & Greenberg, J.M., 1980, Astron.
    Astrophys., 86, L3-6.
Hagen, W. Tielens, A.G.G.M. & Greenberg, J.M., 1981., CHem. Phys., 56
    (3), 367-379.
Hagen, W. Tielens, A.G.G.M. & Greenberg, J.M., 1982, Astron. Astrophys.,
    accepted for publication.
Handbook of Chemistry and Physics, 47th Edition, 1966-1967, The Chemical
    Rubber Co.
d'Hendecourt, L., Allamandola, L.J., Baas, F. & Greenberg, J.M., 1982,
    Astron. Astrophys. Lett., Accepted for publication.
Herbig, G.H., 1975, Astrophys. J., 196, 129-
Hong, S.S. & Greenberg, J.M., 1980, Astron. Astrophys., 88, 194-202.
Van de Hulst, H.C., 1949, Rech. Astr. Obs., Utrecht, 11, part 2.
Van de Hulst, H.C., 1957, Light Scattering by Small Particles, J. Wiley
    & Sons.
Jenkins, E.B. & Shaya, E.J., 1979, Astrophys. J., 231, 55.

Khare, B. & Sagan, C., 1973, In Molecules in the Galactic Environment,
    ed. M.A. Gordon & L.E. Snyder, J. Wiley & Sons, New York.
Kobayashi, Y., Kawara, K., Sato, S. & Okuda, H., 1980, Pub. Ast. Soc.
    Jap., 32, 295.
Kwan, J., 1979, Astrophys. J., 229, 567-577.
Kwok, S., 1980, J. Roy. Astron. Soc. Can., 74, no. 4, 216-233.
Lada, C.J. & Harvey, P.M., 1981, Astrophys. J., 245, 58-65.
Larson, R.B., 1981, Mon. Nat. Roy. Astr. Soc., 194, 809-826.
Léger, A., Klein, J., De Chevergne, S., Guinet, C., Defourneau, D. &
    Belin, M., 1979, Astron. Astrophys., 79, 256-259.
Mann, A.P.C. & Williams, D.A., 1980, Nature, 283, 721-725.
McMillan, R.S., 1978, Astrophys. J., 255, 880-886.
Merrill, K.M., 1934, Pubs. Astron. Soc. Pac., 46, 206.
Merrill, K.M., Russell, R.W. & Soifer, B.T., 1976, Astrophys. J., 207,
    763.
Metzger, P.G., Mathis, J.S. & Panagia, N., 1982, Astron. & Astrophys.,
    105, 372-388.
Morton, D.C., 1974, Ap.J. (Lett.), 193, L35-L39.
Mukai, T., Mukai, S. & Noguchi, K., 1978, Astrophys. Sp. Sci., 53, 77.
Oort, J.H., 1954, Bull. Ast. Inst. Ned., 12, 177-186.
Oort, J.H., 1974, in Recent Radio Studies of Bright Galaxies, ed. J.R.
    Shakeshaft, 375.
Phillips, T.G., Huggins, P.J., Kuiper, T.B.H. & Miller, R;E., 1980,
    Astrophys. J. (Lett.), 238, L103-L.
Purcell, E.M., 1976, Astrophys. J., 206, 685.
Rowan-Robinson, M., 1979, Astrophys. J., 234, 11-128.
Sandell, G. & Mattila, K., 1975, Astron. Astrophys., 42, 357.
Savage, B.D. & Mathis, J.S., 1979, Ann. Rev. Astr. Astrophys., 17, 73,
Schull, J.M. & McKee, C.F., 1979, Astrophys. J., 227, 131.
Schultz, A., Lenzen, R., Schmidt, Th. & Proetel, K., 1981, Astron.
    Astrophys., 95, 94-99.
Scoville, N.Z. & Hersh, K., 1979, Astrophys. J., 29, 578-582.
Serkowski, K., Mathewson, D.S. & Ford, V.L., 1975, Astrophys. J., 196,
    261.
Silverstein, R.M. & Bassler, G.C., 1967, Spectrometric Identification of
    Organic Compounds, J. Wiley & Sons, New York.
Silk, J. & Norman, C., 1980, Interstellar Molecules IAU Symposium no.
    87, ed. B.H. Andrew, D. Reidel, Dordrecht, 165-172.
Smith, D. & Adams, N.G., 1977, Astrophys. J., 217, 741-748.
Spitzer, Jr., L., 1978, Physical Processes in the Interstellar Medium.
    J. Wiley & Sons, New York.
Taff, L. & Savedoff, M., 1972a, MNRAS, 160, 89-97.
Taff, L. & Savedoff, M., 1972b, MNRAS, 164, 357-374.

Tielens, A.G.G.M. & Hagen, W., 1982, Ph.D. Thesis, A.G.G.M. Tielens, Leiden,

Völk, H.J., Jones, F.C., Morfill, G.E. & Roser, S., 1980, Astron. Astrophys., $\underline{85}$, 316-325.

Waters, J.W., Gustincie, J.J., Kaken, R.K., Kuiper, T.B.H., Roscoe, H.K., Swanson, P.N., Rodrigues Kuiper, E.N., Kerr, A.R., & Thaddeus, P., 1980, Astrophys. J., $\underline{235}$, 57.

Watson, W.D., 1976, Rev. Mod. Phys., $\underline{48}$, 513.

Watson, W.D. & Salpeter, E.E., 1972, Astrophys. J., $\underline{174}$, 321.

Whittet, D.C.B., 1981, Q.Jl.R. Astr. Soc., $\underline{22}$, 3-21.

Whittet, D.C.B. & Blades, J.C., 1980, Mon. Not. Roy. Astr. Soc., $\underline{191}$, 309-319

Whittet, D.C.B. & Van Breda, I.G., 1978, Astron. Astrophys., $\underline{66}$, 57.

Whittet, D.C.B., Bode, M.F., Evans, A. & Butchart, I., 1981, Mon. Not. R. Astr. Soc., $\underline{196}$, 81P-85P.

Wickramasinghe, D.T. & Allen, D.A., 1980, Nature, $\underline{287}$, 518-519.

Willner, S.P., Puetter, R.C., Russell, R.W. & Soifer, B.T., 1979, Astrophys. Sp. Sci., $\underline{65}$, 95-101.

Willner, S.P., Gillett, F.C., Herter, T.L., Jones, B., Krassner, J., Merrill, K.M., Pipher, J.L., Puetter, R.C., Rudy, R.J., Russell, R.W. & Soifer, B.T., 1982, Astrophys. J., $\underline{253}$, 174-187.

Willner, S.P. & Pipher, J.C., 1982, Preprint, Proceedings of Workshop on the Galactic Center, California Institute of Technology, Jan. 1982.

Woodward, P.R., 1978, Annual Rev. Astron. Astrophys., $\underline{16}$, 555-584.

Wootten, A., Evans, N.J. II, Snell, R. & Vanden Bout, P., 1978, Astrophys. J., $\underline{255}$, L143-L148.

SECTION IV

Submillimetre wave instrumentation

BAND-PASS FILTERS FOR SUBMILLIMETRE ASTRONOMY

S. T. Chase
Imperial College of Science and Technology

R. D. Joseph,
Imperial College of Science and Technology

INTRODUCTION

The construction of a wideband mm-wave photometer requires efficient filters fitting the atmospheric transmission bands. The 'black art' of infrared filter design generally involves the use of crystals, plastics, or even cardboard in order to construct a suitable pass-band. In contrast, periodic metallic arrays offer designable filters for wavelengths greater than 100μ. An array of square apertures in a conducting film performs as a high pass filter element, and its complementary structure of conducting square acts as a low pass element (Ulrich 1968). An array of cross-shaped apertures, by contrast, exhibits a band-pass behaviour, and such arrays are the subject of this paper.

ARRAY CONSTRUCTION

Arrays of crosses with five different shapes were fabricated. The shape parameters varied were cross arm length L, arm width 2b, and cross separation 2a, as shown in Fig. 1. The array periodicities g = L - 2a ranged from 330μ to 550μ.

The arrays were fabricated by standard photo-etching techniques. Both free-standing arrays etched in 4μ nickel foil and supported arrays, etched in an aluminium film deposited on a thin Mylar substrate, were produced. It was possible to achieve good quality, consistently, in the etched pattern for the supported arrays, whereas the quality of the free-standing arrays was generally poorer, and they were very fragile.

SPECTRAL PERFORMANCE

Typical examples of transmission spectra for array types 1 to 3 are shown in Fig. 2, and for array types 4 and 5 in Fig. 3. The general features of these five spectra demonstrate that variations in the cross shape parameters do affect the peak transmission wavelength

λ_m, the bandwidth $\delta\lambda/\lambda_m$, the percentage peak transmission T_m, and the short wave rejection.

The resonant wavelength is determined principally by the cross arm length L, and not by the array period, g, as has usually been assumed in the literature.   In Fig. 4 a plot of λ_m/L vs a/L demonstrates that for a/L >   0.15, λ_m = 2L to within   10%.

The filter bandwidth can be varied dramatically by altering the cross separation in the array, and for small cross separations, large bandwidths may be obtained.   This is illustrated by the plot

Figure 1

a)   Five types of master pattern used to generate arrays.
b)   Definition of array parameters: g = array period, L = cross-arm length,   2a = cross separation, 2b = cross arm width.

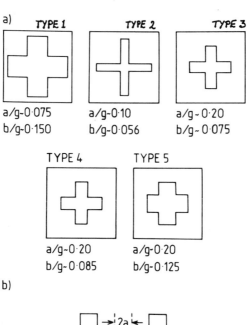

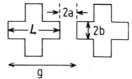

of $\delta\lambda/\lambda_m$ vs a/L in fig. 5. The curve drawn through the data in Fig. 5, $\delta\lambda/\lambda_m = 0.14 + 0.013\ (a/L)^{-3/2}$, should provide a good guide for designing a filter of a specified bandwidth, although similar curves proportional to $(a/L)^{-v}$, with $1 < v < 2$ also fit the data reasonably well.

The peak transmission and shortwave rejection (1st and 2nd harmonics) are critically-dependent on the array quality. For all give types the peak transmission can exceed 90%. With the exception of

Figure 2

Transmission spectra for array types, 1, 2 and 3, with g = 410µ. All arrays are aluminium on 10µ Mylar substrate.

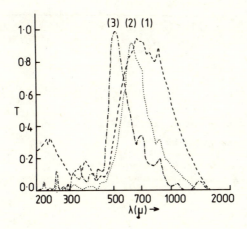

Figure 3

Transmission spectra for array types 4 and 5 with g ≃ 375µ. Also aluminium on Mylar.

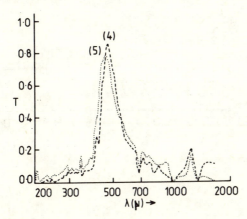

Figure 4

λmax/L vs a/L for a sample of 24 arrays. Empirical
curve is shown superimposed.

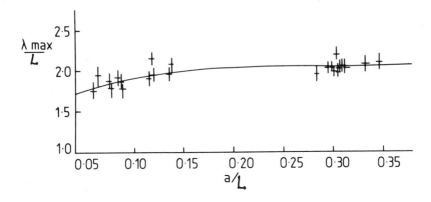

Figure 5

$\delta\lambda/\lambda_{max}$ vs a/L for some sample of arrays. Curve shown
superimposed is given by $\delta\lambda/\lambda_{max} = 0.013(^a/L)^{-3/2} + 0.14$.

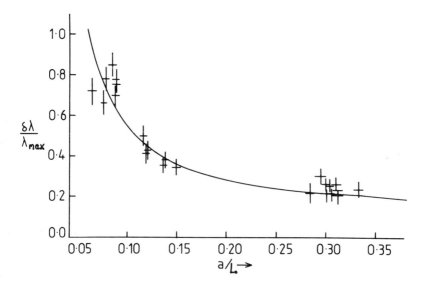

type 1, all other shapes exhibit only narrow transmission features with T ∿5-10% at the first two harmonics. Type 1, however, has shortwave rejection two-to-three times less effective, as might be expected from its "fat" shape. Defects in the quality of the reproduced cross pattern invariably result in reduced peak transmission, poor shortwave rejection, and increased bandwidth and resonant wavelength.

CONCLUSIONS

Resonant arrays of simple crosses offer an attractive solution to the problem of filter design for far-infrared and sub-millimetre astronomy. They offer excellent peak transmission combined with good short wave rejection. Using the results presented above they may be designed to provide transmission filters of the wavelength and bandpass required to fit the various atmospheric windows, or for other applications.

ACKNOWLEDGEMENTS

It is a pleasure to thank Dr. Peter Ade who first brought the resonant array idea to our attention, and Mr. Richard Chater whose help with the FIR spectroscopic measurements was invaluable. Thanks also to Lucinda Symons and Nick Jackson for their assistance in preparing the photographic masks for which all of our arrays were generated. S.T.C. holds a research studentship from the U.K. SERC.

REFERENCES

Anderson I. (1975) On The Theory of Self-Resonant Grids. The Bell System Tech. J., 54, no.10 1725-1731

Arnaud J. A. & Pellow F. A. (1975) Resonant-Grid Quasi-Optical Diplexers. The Bell System Tech.J., 54 no.2 263-283

Davis, J. E. (1980) Bandpass Interference Filters for Very Far Infrared Astronomy. Infrared Physics, 20 287-290

Ulrich, R. (1967) Far Infrared Properties of Metallic Mesh and its Complementary Structure. Infrared Physics, 7 37-55.

_____ (1968) Interference Filters for the Far Infrared Applied Optics, 7 no.10 1987-1996

_____ (1969) Preparation of Grids for Far Infrared Filters. Applied Optics 8 no.2 319-322.

THE SUBMILLIMETER RECEIVER OF THE FUTURE

P.Encrenaz
Ecole Normale Supérieure   and   Observatoire de Paris

Abstract. An investigation of the different types of receivers used today at millimeter and submillimeter wavelength is carried out. Forecasts for their behavior at 300 GHz and above for the next five years are presented.

We will limit ourselves to the wavelength range $700\mu \leqslant \lambda \leqslant 3mm$ for ground-based observations. As it can be seen from Fig.1, the contribution from the atmosphere to the system temperature T_s is always greater than a few tens of Kelvin, therefore pushing the radioastronomer to have a receiver whose T_s is sky noise limited, and not receiver limited.

Every observer wishes a receiver having all the following qualities:

- tunable
- broadband
- sensitive
- large dynamic range
- as little local oscillator power as possible.

As always a compromise is necessary, and the goal of a sky noise limited receiver demands the best available sites (Hawaii-Mauna Kea, Kitt Peak, Plateau de Bures...), implying a reliable receiver.

I **The Resistive Shottky Mixer**

The noise figure of such a receiver can be expressed as

$$T_R = T_M + L\, T_{IF}$$

where T_M = mixer temperature (°K)

L   = conversion loss (dB)

T_{IF}= IF noise contribution (°K).

The non-linear element is a Shottky diode whose characteristics
are typically:

$$C_o = 3 - 10 \text{ fF}$$
$$R_s = 5 - 8\,\Omega$$

giving $\nu_{cut-off} \simeq$ few T Hz.

In principle, such a device is usable at all mm and submm
wavelengths. However, its matching to the waveguide structure, the losses
in the waveguide, the need of an injection cavity have been strong limitations to its good performances at short wavelength.

Major improvements have been implemented over the last 12 years
for the Shottky mixer:

- Cooling of the mixer to 20 K in most cases: the classical
Nyquist formula $\langle e^2 \rangle$ = 4kRBB requires us to cool the resistive mixer to a
temperature as low as possible (k=Boltzmann constant,Joule K^{-1}; R=resis-
tance,ohm; B=bandwith,Hz; T=thermodynamic temperature of R; $\langle e^2 \rangle$ =mean
quadratic voltage) . Routine operation of Shottky mixers at 20K has taken
place in 6 observatories over the last 8 years. Blum, Kerr and Weinreb
have pioneered this field.

- Physics of the mixer: impedence matching of the Shottky
junction in its reduced height waveguide at harmonic frequencies of the
local oscillator (L.O.) and signal frequencies has improved tremendously
the performances of the mixer (Held, Kerr,...).

- Molecular Beam Epitaxied AsGa has resulted in much better
characteristics for the Shottky junction, and a need for 5 to 10 times
less L.O. power. Mattauch, Schneider and Wrixon have produced diodes
where T_M is an order of magnitude lower than in early mixers.

- New FET amplifiers have greatly enhanced the reliability of
the system.

As to the IF stage, the technology of submicron lines in FET's (mostly Mitsubiski 1402 and 1412 transistors), and the cooling of the AsGa FET's give parametric amplifiers the status of historical monsters in the 1 - 5 GHz band. Weinreb has designed simple circuits which are used worldwide. T_{IF} varies from 7 to 15K from 1 to 5 GHz with an IF bandwidth of 500 MHz.

- Quasi optical injection schemes have permitted operation well below 1 mm. However few tests have been carried out at low temperature.

- Doublers and triplers with high efficiency are becoming increasingly common: diodes with large back-breakdown voltage are now available, and permit a few milliwats of L.O. power even at 350GHz. Archer has to be credited for these major progresses.

The present and expected best performances are (SSB):

		3 mm	1.5 mm	.9 mm
1981	T_R (K)	180	300	1000
1985	T_R (K)	60–100	150–200	500?

II The InSb hot electron bolometer

A major effort has been done here at QMC with this system, and at Cal Tech.

At liquid He temperatures, there are still electrons in the conduction band which are not frozen out. A small current in the bulk material will raise T_e, the electron gas temperature, considerably. The electron mobility is a strong function of T_e. The energy relaxation time of the electron gas is $\sim 10^{-7}$s. This will limit the I.F. passband to a few MHz for a bolometer mounted in a waveguide (it acts as a mixer, in that it responds to the square of the electric field at that point in the guide).

Major progress has been made in the I.F. amplifiers (5K or less for 500 Ω input) and in high frequency observations (White, Beckman, Phillips...).

Major astronomical observations have been made from 3 to .5 mm with such systems using the scanning in frequency or the chopping on the sky. The new generation of dewars (possibly to He3,Coron,Chanin...) should help to improve the performances of InSb bolometers (conversion loss is ~ 10 dB, and $T_{IF} \sim 5K$). The limitation is the bandwidth .

		3 mm	1.5 mm	.5 mm
1981	T_R	250	250	600
1985	T_R	100?	100?	300?

III The SIS Mixer

The direct use of the Josephson effect has been relatively deceptive, although major efforts have been made in this direction (Taur, Kerr...). Major progress in SIS junctions, and their recent availability has allowed radioastronomers to build receivers at 35 and 115 GHz.

Progress in understanding the physics of the mixer has also been very rapid (Tucker, Richards, Feldman, Kollberg...). However only the low IF limit is understood, and $h\nu/k \ll \Delta$ (gap of the supraconductor). Gain is predicted, and has been effectively observed at 35 GHz.

Some results have been obtained with arrays of junctions in series. The use of arrays of junctions lessened the difficulty of fabricating extremely small area junctions (.1 to .5 square μ). $T_M \propto n^{\beta}$, where $\beta = 1$ or -0.5 (?), n being the number of junctions in series.

The potential advantages of such mixers have pushed more than 20 groups in the world to work in this direction. Some of the remaining problems are:

- how to avoid the Josephson noise (Meissner effect)?
- is gain possible at submm wavelength with SIN junctions?
- how to work out a noise theory for SIS and SIN mixers?

		8 mm	4 mm	2.6 mm	1.3 mm	1 mm
	L	gain 5 dB	2 dB	7 dB	10–12 dB	–
1981	T_R	$\leqslant$10 K	70 K	150 K	600 K ?	–
1985	T_R	2.3 K ?	5 K ?	8 K ?	15 K ?	gain?

IV Rydberg Masers

For a Rydberg atom with principal and azimuthal quantum numbers n and l respectively, classical theory applies for $n \gg 20$. On a circular orbit:

$$r_n = n^2 a_o \quad , \qquad v_n = v_o/n$$

with $\qquad a_o = .5 \overset{o}{A} \quad , \qquad v_o = 2.3 \ 10^6 \ km \ s^{-1}$.

With a small value of l, such an atom has a huge dipole moment; the excitation of alkali atoms with dye lasers does allow population of a chosen sublevel n with a high efficiency (Haroche, Kleppner, Walter), hence a maser effect between the n and n–1 levels, which has been observed at millimeter wavelengths.

An equivalent N.E.P. at 107 GHz of $\sim 10^{-16}$ W Hz$^{-1/2}$ has been obtained with an uncooled system, while $\sim 10^{-18}$ W Hz$^{-1/2}$ is obtained with a cooled system.

In principle the device is narrow-band, but Stark-effect or double irradiation broadening are easy to achieve.

This system has been used as a preamplifier at $\lambda = 3$ mm, and lowered by a factor 40 the noise of a classical Shottky mixer over a 4 MHz bandwidth. Its other advantage is the easiness to reach 1000 GHz, with the same basic equipment.

This system is potentially extremely attractive for high frequency, and suffers only from its large present dimensions. It is difficult to put it in the focal box of an antenna, unless one has a Coudé design.

V Other types of receivers

Down converters and subharmonically pumped mixers have been deceptive in the sense that their performances do not match those of a single Shottky mixer.

A Ge:Sb photoconductor has been recently used at .1 mm to carry out heterodyne spectroscopy.

VI Conclusions

Major efforts are necessary for SIS and Rydberg masers to compete with InSb performances below 1 mm, but there is hope to have a sky noise limited receiver at 1 mm before the end of the decade.

References

- Review Papers:

  - BLUM,E.J. Radioastronomy at millimeter wavelength in "Advances in Electronics and Electron Physics",56,97
  - PHILIPS,T.G. and WOODY,D.P. Millimeter and Submillimeter-wave receivers. 1981,preprint n°25

- We also mention the following papers, not included in the two previous review papers:

  - BEAUDIN,G.,LAZAREFF,B. and MAHIEU,J. 115 GHz low noise cryogenics receiver for radioastronomy . Preprint, Observatoire deeParis

  - BAUDRY et al . A new spectroscopic facility at Millimeter wavelengths. J.Astrophys. Astr. 1,193

  - FABRE,C. Une étude des états de Rydberg en ondes millimétriques; thèse d'Etat, Laboratoire de Physique, Ecole Normale Supé.÷ rieure,1980.

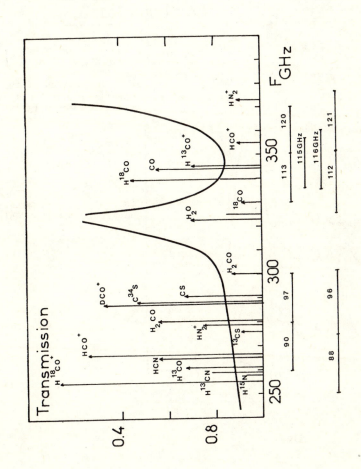

Fig. 1 : Atmospheric transmission near 300 GHz in a high altitude site. At 350 GHz, Tsky will be around 50 K. The heavy lines indicate the frequency coverage obtained with a tripler(and a klystron as L.O.).

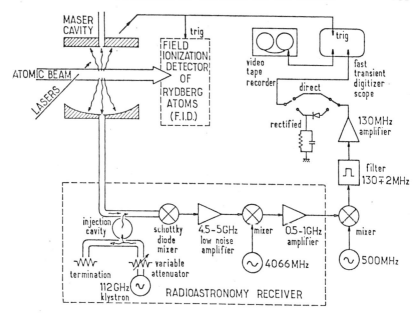

FIGURE 2

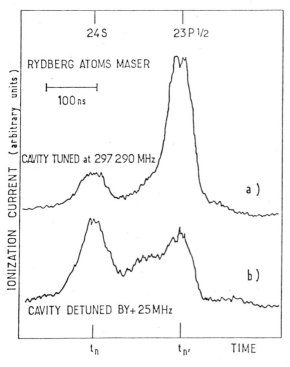

Fig. 2 : sketch of the experiment which has permitted the first detection of a Rydberg maser at 107 GHz.

Fig. 3 : The tuning of the cavity to the exact frequency of the transition (245 → 23 P 1/2) greatly increases the transfer of population of the sublevels.

FIGURE 3

HIGH FREQUENCY TECHNIQUES IN HETERODYNE ASTRONOMY

Thijs de Graauw, Astronomy Division, Space Science Department of ESA, Noordwijk, The Netherlands

Abstract. This paper describes the advances in several techniques for heterodyne detection at frequencies above 200 GHz. Local oscillators and mixers in use for astronomical observations and current developments in these areas are reviewed.

I. Introduction

In the past decade most of our understanding of the interstellar and circumstellar medium has been obtained through observations of molecular emission and absorption lines, mainly at millimeter wavelengths. For the instrumentation at these wavelengths heterodyne techniques have been used which are very similar to those applied at cm wavelengths, although with overall component dimensions scaled down linearly with wavelength.

The principles of heterodyne detection are well known these days but will be summarized again here. Radiation from a telescope or antenna is combined, through an optical or waveguide device, with monochromatic radiation from a local oscillator into a non-linear mixing element. Output at the difference freqency is amplified and subsequently fed, often after another downconversion step, into a backend spectrometer. The aimed for high spectral resolution is then obtained by utilising a bank of frequency selective radiofilters, or other dispersive techniques.

The interesting astronomical information on the interstellar medium collected in the mm region argues for observations at even shorter wavelengths. This is clearly illustrated in most of the contributions to this conference. However, development of techniques for the short millimeter-submillimeter range has been very slow. The main reason for this is that there are not many fields in science that require sensitive high frequency heterodyne receivers. Besides astronomy and atmospheric physics, the main support for technological improvements comes from

plasma diagnostics, where the studies of the plasma state have been
oriented towards the achievement of controlled thermonuclear fusion (see
Luhmann, 1979).

Nevertheless several advances have recently been made, and it
appears that state-of-the-art submillimeter instrumentation has now moved
into the linear part of the curve that describes the relation between
effort (manpower, money) and performance (sensitivity, bandwidth etc;
see figure 1).

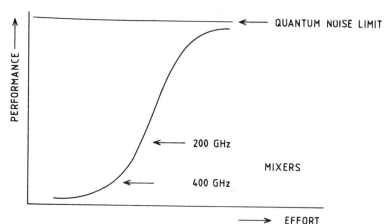

Figure 1 – Relation between effort (eg. manpower) and performance (sensitivity) for mm-submm receivers in 1981.

With the advent of high frequency receivers the need for larger and highly accurate antennas has become clear. Until now all submillimeter observations were carried out with optical and infrared telescopes,
which have relatively small apertures and are often not as efficient as
radio antennae used to be. The availability of these telescopes to submillimeter astronomers is however very limited, since they are competing
with a large and strong community for observing time. Progress in antenna
design and technology, often involving radio astronomers themselves, has
led to the planning and, in several cases, to the construction of large
(> 10 m) telescopes with a surface accuracy that will allow them to be
used at frequencies as high as 500 GHz, and in one case even up to 1 THz.
Some of these telescopes will be located at the best mountaintop site
available for astronomy, Mauna Kea, Hawaii, where the atmospheric transmission is often high enough to allow observing up to 400 GHz, and some-

times even up to 1 THz. These telescopes will not only require receivers
with a high sensitivity, but also large IF system and backend instantan-
eous bandwidths. For example, in order to obtain a 500 km s^{-1} velocity
coverage, necessary for observations of the galactic center and nuclei of
other galaxies, the required bandwidth is 500 MHz at 1 mm, and 1 THz at
0.5 mm wavelength.

   The complicated and very diverse technologies involved in a
submillimeter receiver make them difficult to use by the average astronomer. Together with the fact that facilities must preferably be located at
high altitudes where working conditions are difficult, a certain degree
of automation, and the capability of complete remote control by radio
link, will become indispensable for efficient and successful observing.

   In this paper we will review local oscillators and super-
conducting mixers for short millimeter and submillimeter wavelength
receiver instrumentation. We will also give a survey of receivers
presently operating at frequencies above 200 GHz.

II. Local Oscillators

   Lack of sufficient local oscillator power to drive mixers has
often impeded developmental programmes for high frequency receivers.
Two lines of research have changed the local oscillator situation for
frequencies above 200 GHz. First of all, several advances in Schottky
diode fabrication have relaxed considerably the requirements for the l.o.
output power level. This now allows us to use frequency multipliers
together with solid state oscillators and klytrons, that can deliver
power levels in excess of 1 mWatt up to 350 GHz. Secondly, improvements
of several technologies necessary for the fabrication of backward wave
oscillators (b.w.o.) have resulted in considerable improvements in the
characteristics of these monochromatic sources. These tubes, which offer
a 20% tuning range at 500 GHz, are currently available, and can deliver
25 mWatts of power at the fundamental frequency. Tubes with equivalent
capabilities at 1 THz are presently under construction.

   A third type of L.O. source used in submillimeter receivers
is the optically pumped molecular gas laser. These lasers have the
advantage of providing output powers (> 1 mWatt) over the entire far-
infrared and submillimeter range, but their discrete tuning property is a
serious limitation in their use, since a close coincidence between laser
lines and interstellar line frequencies is necessary. Molecular lasers

use a 10 μm CO_2 pumplaser with a high voltage supply, both mounted on a stable optical table. The instrument is therefore rather clumsy and has only been used in a fixed (Coudé) focus arrangement.

Attempts are presently being made to construct a system involving a moving focus arrangement by Betz. An example of a laser localocal oscillator used in observations of interstellar CO(6-5) emission is given in figure 2.

SUBMM HETERODYNE SYSTEM

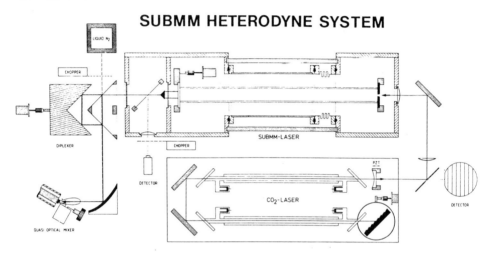

Figure 2.   Submm Heterodyne system with a laser local oscillator after Schultz 1980.

An overview of operational local oscillator sources and their operating range for diode mixers and InSb mixers is given in table 1. They will be discussed in detail in the following sections.

Backward wave oscillators

Detailed descriptions of BWO's have been given by several authors (Convert and Yeou, 1964; Heffner, 1954; Kantorowicz and Palluel, 1979). BWO's or carcinotrons (the trade name for BWO's manufactured by Thomson CSF) were invented in 1951 in two models, one using crossed fields (M type), the other using parallel fields (O type). The first tube operating at mm wavelengths was constructed in 1953. Developments in 1975 resulted in a tube operating at 1.5 mm and the 1 mm barrier was crossed in 1960. The shortest wavelength tube ever operated was at

System	Frequency Range	
	∿ 1 mWatt (diode mixer)	∿ 1 μWatt (InSb mixer)
Klystron + Multiplier	< 350 GHz	≤ 500 GHz
Solid State Osc. + multiplier	< 290 GHz	
BWO	≤ 530 GHz	
BWO + doubler		≤ 625 GHz
Molecular laser	∿ 700 GHz	

Table 1. Local Oscillator systems used in mm/submm receivers.

around 0.25 mm in 1969. We deal here with O-type carcinotrons only.
M-type oscillators are intrinsically more noisy than the O-type, which
have been shown to be excellent coherent oscillators at submillimeter
wavelengths.

    Operation - In a BWO an electron beam, generated by an elect-
ron gun, is propagated through a slow-wave structure (delay line) along
which an electro-magnetic wave is also propagating. A strong interaction
occurs when the phase-velocity of the wave is close to the electron
velocity, which is in turn modulated by the longitudinal component of the
slow-wave electric field. Part of the electron beam kinetic energy is
then exchanged with the energy of the electromagnetic wave travelling in
the opposite direction. The electron gun and the delay line are the
principal elements of a BWO, and represent the most difficult elements to
construct. In the electron gun the filament heats the cathode material
to a temperature that ensures the required electron emission density.
The electron beam is subsequently focussed into the delay line by a multi
element electrode (Wehnelt and anode). Because of the small component
dimensions involved, focussing of the beam into the delay line is very
critical. The electron beam is guided along the delay line by a magnetic
field and its velocity is controlled by the potential difference between

the delay line and the electron gun. The electromagnetic beam is extract-
ed from the delay line near the electron gun.

Figure 3a, b, c shows various details in the contruction of
carcinotrons. The relative dimensions of the vanes in the delay line
are given in figure 3c. For a tube oscillating at 500 GHz, the actual
dimensions are obtained by multiplying the numbers by 50 μm.

Characteristics

The use of BWO as local oscillators in low noise receivers
started only after 1977. For reasons which are not entirely clear, these
sources were considered to be very noisy, and thus unsuitable for receivers. They were mainly used in molecular spectroscopy experiments
(Krupnov and Gershtein, 1970) and in solid state research (Tuchendler et
al. 1973).

During 1976 and 1977 experiments showed that the noise at
1.4 GHz from the carrier (at 240 GHz) amounted to about 120 db/MHz, and
contributions of this local oscillator to the radiometer noise temperature
were between 1000 K and 3000 K when no diplexer was used (de Graauw et al.
1978). With a 20 db rejection from the diplexer this figure becomes
negligible for today's receivers. These findings are in agreement with
previous X-band experiments reported by MacKenzie et al. (1964).
Difficult access to this report may have prevented a more widespread use
of BWO's. Another series of measurements with BWO's as local oscillator,
now with InSb bolometer mixers instead of diode mixers, also showed
evidence of its low noise characteristics (van Vliet et al. 1979).
Further experiments near 462 GHz indicated that the noise level can
increase dramatically near r.f. power dips (van Vliet et al. 1982). This
is shown in figure 4 where the sideband noise is plotted as a function of
frequency, and where the power-dip is given by the dotted line. The sideband noise was measured with constant power absorbed in the InSb mixer to
keep the conversion efficiency constant. This prevented the use of this
tube for observations of CO(4-3) emission even with sufficient L.O. power
available to drive the mixer. A complete frequency to output power curve
for this tube is given in figure 5c. Other reports on BWO noise indicated
in one case an increase of noise with IF frequency, in another an apparent sideband noise at 4 GHz of about 20,000 K. With a tube at the end of
its lifetime, caused by the breakdown of the filament, we observed an
increase in the single sideband noise temperature of the receiver of 100 K

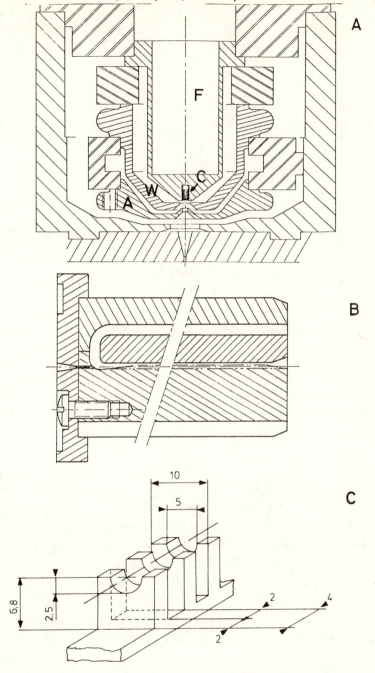

Figure 3.   Details of the construction of a BWO. (a) shows the electron
gun with the filament F and cathode material C, the focussing
elements wehnelt W and the anode A. (b) shows the delay cir-
cuit with r.f. output coupling structure. (c) shows the
relative dimensions of the vanes of the delay line.

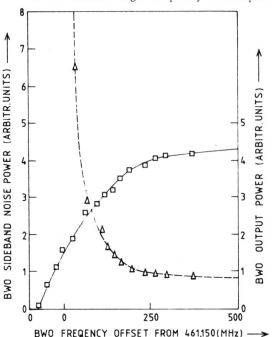

Figure 4.   Increase of B.W.O. noise in a region of strongly decreasing
             output power.

per day, which is equivalent to $\sim$ 800 K per day increase of BWO noise.
In general one can conclude that BWO noise characteristics are quite
acceptable for low noise receivers in astronomy (Lidholm and de Graauw,
1979), atmospheric physics (Waters et al. 1979) and plasma diagnostics
(Luhmann, 1979).

Recent developments - In 1975 the European Space Agency
initiated a research and development program   on BWO's to be carried out
by Thomson CSF. Almost all aspects of carcinotrons were evaluated and
studied, resulting in a step by step improvement of overall performance,
and many other practical aspects. Table II gives a summary of the
improvements which have been made to date, and we compare those with the
situation in 1975 before the program was started. Figures 5a and 5c show
the output power versus frequency curves of tubes produced in 1975.
Figures 5b and 5d show the corresponding results for 1981. Best progress
has been made in tuning range, output power level at high frequencies, and
lifetime, which is important from the point of view of costs. The latter
improvement was achieved redesigning the filament (potting) and using
advanced cathode materials that allowed lower operating temperatures.

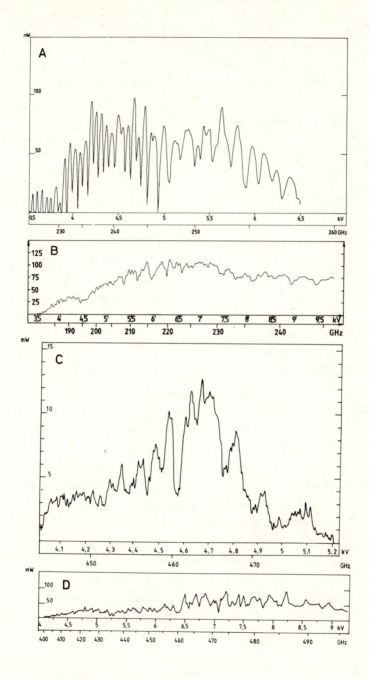

Figure 5. Output power versus frequency curve of BWO's. (a) and (c) produced in 1975. (b) and (d) produced with advanced technology.

BWO characteristics	1975	1981
Tuning bandwidth	10%	20%
output power (> 400 GHz)	5 mW	50 mW
lifetime	< 2000 hrs	8000 hrs
		(expected)
power consumption	250 Watts	< 150 Watts
weight	35 kg	9 kg
bulk	15 l	4 l

Table II - Characteristics of BWO's produced in 1975 and in
1980.

Higher output powers at high frequencies (> 300 GHz) were obtained by
improving the mechanical assembly and the output waveguide. Very recently
the coupling between the delay line and the oversized waveguide has been
changed to avoid the generation of other modes. One problem that has
been noticed in using carcinotrons is the change in direction of the output beam. A variation in frequency by about 5 GHz already changes the
center of the beam about 1° in direction. This may be caused by the
presence of several other modes in the output beam besides the fundamental
mode. Experiments with tapered transitions to the fundamental waveguide
showed a decrease by a factor two in output power. As a first step to
overcoming the skewing problem, the coupling between the delay line and
the output waveguide has been changed. Until recently the coupling guide
itself was oversized. Now the coupling has the dimensions of the fundamental guide at the operating frequency, and a taper is used further-on
to go to the usual oversized output waveguide (RG138/U). Experiments are
in progress to detect beam skewing for the new configuration. The improved output coupling may also have taken care of the large dips in the output power curve in figure 5a, which is absent in figure 5b. Presently
BWO's are commercially available with the improvements noted above. As a
next step in the development programme, a tube is under construction for
operation between 850 GHz and 1000 GHz. It is expected that a 10 mWatt
output power level will be reached over at least 50% of this frequency
range with existing fabrication techniques.

Frequency Multipliers

Advances in the development of frequency doublers and triplers over the past two years allows the use of solid state oscillators which are much less expensive and bulky, and which do not require sophisticated power supplies. Archer (1981 a, b) developed a tripler that exhibits a typical conversion efficiency greater than 5% when tuned to output frequencies between 210 and 240 GHz, and input power levels between 45 and 55 mWatt. These power levels are obtained from a specially designed Gunn oscillator using a commercial Gunn diode in a reduced height waveguide. The oscillator is tuned by varying the position of a contacting backshort. In order to obtain L.O. power above 300 GHz without using solid state oscillators, a two step scheme is now being worked out with one Gunn oscillator and two doublers (Archer, 1981). This scheme allows one to use the Gunn diodes with different characteristics in an optimum way. Triplers with klystron pumps have been made to operate at 350 GHz (Erickson, 1981), but due to the high cost of klystrons they do not represent a viable alternative to carcinotrons.

III. Mixers at high frequencies

The most critical component of a receiver is the mixer. There are several types of mixing elements: diode, photoconductive and super-conductive mixers. They are all used in the short millimetre and sub-millimetre wavelength range. Several reviews have been given on diode mixers (Kelly and Wrixon, 1980) and on InSb photoconductive bolometer mixers (Phillips and Jefferts, 1973). Superconductive mixers have been reviewed by Richards and Shen (1980). They reviewed Josephson Junction (JJ) effect mixers as well as superconductor - insulating oxide - superconductor (SIS) devices. Many groups have attempted to produce stable JJ mixers, with little success. It is therefore not surprising that with the advent of thin-film SIS devices many groups have decided to apply the quasi-particle tunneling devices in their mixers. The technology of these devices is well developed and lifetime and thermal cycling effects have been studied in detail. Junction degradation is also well under control. Excellent results have been obtained up to 115 GHz, and table III reviews the laboratory results to date. Application of SIS devices at a frequency around 230 GHz made the high frequency limitation of these devices very clear (Phillips et al. 1981). Operation of quasi-particle mixers at

	Frequency GHz	T mixer K
Phillips et al. (1981)	115 GHz	140
	230	300
Kerr et al. (1981)	115	40 $\pm$ 50
Lenke et al. (1981)	115	80
Rudner et al. (1981)	75	< 100

Table III - SIS mixer performance.

higher frequencies moves the optimum dc voltage bias point towards lower voltages into the Josephson effect noise region. One potential way to avoid this problem is to apply a magnetic field to reduce the critical current, and this has been done successfully, as shown in figure 6. In experiments by Kerr et al. the Josephson effect noise has apparently been eliminated with even more success (see figure 7 curves c and d). In this experiment an array of 14 junctions have been used resulting for the same impedance, in an arrayed junction with larger area, to which a magnetic field can be applied more efficiently. Whether the Josephson effect noise is a fundamental limitation for frequencies above 300 GHz is not clear. Though higher frequency operation requires smaller RC products and smaller areas to keep ωRC close to unity or lower, experiments by Kerr (1981) indicate that ωRC products larger than unity can also be used.

Josephson junction mixer experiments have been carried out with adjustable point-contact junctions. These point-contacts are not however reproducible. Poorter (1978) has succeeded in producing rather stable point contact junctions for operation at 200 GHz. Table III summarizes the results of Josephson Junction mixer experiments above 100 GHz. The results of Daalmans et al. were obtained with evaporated thin film Nb-No junctions made with a technology analogous to that of SIS device fabrication. They are thermally recyclable and have showed no degradation when exposed to the ambient environment, which is not the case with Pb SIS junctions. Although fabrication requires more sophisticated evaporation equipment, several SIS production groups are now trying to master the Nb evaporation process.

In principle, Josephson Junctions can be used up to very high frequencies and their characteristic frequency is given by the energy gap in the junction material, as well as the operating temperature. These

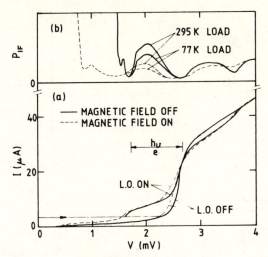

Figure 6.   (a) Current vs. voltage characteristics for a junction with
             and without 230 GHz LO power applied.   (b) IF output with hot
             and cold load signals applied.   The effect of applying a mag-
             netic field is shown by the dashed lines.   After Phillips et
             al. 1981.

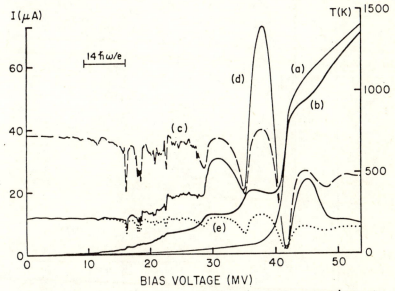

Figure 7.   Experimental results for a 14-junction SIS array mixer at
             115 GHz.   (a) The unpumped DC I-V curve.   (b) The DC I-V
             curve with LO power P_{LO} = 0.375 μW, showing a region of
             negative differential resistance.   Curves (c)-(e) show the
             IF radiometer output with (c) a noise source applied to the
             mixer's IF port, (d) a monochromatic source applied to the
             mixer's signal port, and (e) no signal or IF power applied;
             all with P_{LO} = 0.375 μW.   The array normal resistance (R_A)
             measured from (a) is 600 Ω.   Kerr et al. (1981).

Frequency	T_mSSB K	Conversionloss (dB)	Reference
135	180	5	Richards
185	165	4.5	Poorter
230	380	6	Daalmans
300	223	9.5	Edrich
450	350	7.7	Blaney

Table III - Experimental results of Josephson Junction mixer experiments.

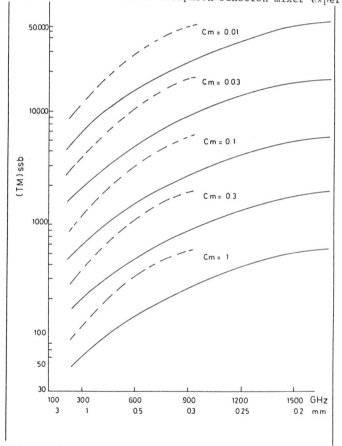

Figure 8.   Mixer noise temperature for N_b - N_b and N_{b_3}Sn - N_{b_3}Sn junctions, operating at 5k, for various r.f. input coupling efficiencies (C_m), as a function of frequency. After Blaney and Cross.

frequencies are 1140 GHz for Nb and 2100 GHz for Nb_3Sn if they are used well below their critical temperature. Blaney and Cross (1978) have calculated mixer performance for these two materials for various r.f. input coupling efficiencies (C_m) as a function of frequency (see figure 8). It is clear from research and development on Josephson Junctions that these devices still represent an interesting way to obtain low noise receivers at very high frequencies.

IV. Receivers operating above 200 GHz

In table V we give an overview of receivers operating above 200 GHz. Several of these receivers are now in use at various observatories with optical telescopes.

	Organisation	Frequency Range GHz	Observatory
DIODE	NRAO	220 - 290	Kitt Peak
	Univ. of Texas	220 - 290	MWO
	Univ. of Mass.	230 - 350	MWO/MMT
	MPI f R	210 - 240	ESO
	ESTEC/Utrecht	220 - 400	ESO/IKIRT/C-141/IRTF
	QMC/MRAO/Kent	220 - 260	UKIRT
	MPI f R (laser l.o.)	690	JungFrau Joch
	FCRAO (laser l.o.)	690	IRTF
InSb	QMC	220 - 350	UKIRT/CARSO
	Caltech	220 - 500	OVRO/C-141
	MPI f R	460 - 500	Mt. Lemmon

Table V - Receivers operating above 200 GHz, in use for astronomical observations.

References
Archer, J.W., IEEE Trans MTT-29, 552, 1981a.
Archer, J.W., 6th International Conference on IR and MM Waves Miami, 1981b.
Archer, J.W., Private communication, 1981c.
Blaney, T.G. and Cross, N., NPL report S.I. 89/0382, 1978.
Blaney, T.G., Future Trends in Superconductive electronics. Edts: Deaver et al., New York, AIP conference Proc. 44, 230, 1978.
Claassen, J.H. and Richards, P.L., J. Appl. Phys. 49, 4130, 1978.

Convert, G. and Yeou, T., Millimetre and Submillimetre Waves, chap. 4. Benson ed. Iliffe Book, London, 1964.

Daalmans, G.M., de Graauw, Th., Lidholm, S., and van Vliet, F., SQUID, Berlin 1980.

Edrich, J., 4th Int. Conf. on IR and Submillimetre Waves. Perkowitz editor. Miami, 1978.

Erickson, N.R., Private communication, 1981 .

Fetterman, H.R., Tannenwald, P.E., Clifton, B.J., Parker, C.D., Fitzgerald, W.D., Erickson, N.R., Appl. Phys. Lett. 33, 151, 1978.

de Graauw, Th., Anderegg, M., Fitton, B., Bonnefoy, R., Gustencic, J.J., 3rd Int. Conf. on Submillimetre Waves, Guildford, March 1978.

Heffner, H., Proc. IRE, 42, 6, 930, 1954.

Kantorowicz, G., Palluel, P., Infrared and Millimetre Waves, vol. 1. K. Button, editor Academic Press, 1979.

Kelly, W.M., Wrixon, G.T., JR and MM Waves, vol. 3. K. Button editor Academic Press, 1980.

Kerr, A.R., Pan, S.-K., Feldman, M.J., Davidson, A., Physica 108B, 1369, 1981.

Krupnov, A.F., Gershtein, L.I., Pribory i Tekh. Eksperim., 6, 143, 1970.

Lidholm, S., de Graauw, Th., 4th Int. Conf. on IR and Submm Waves, Miami, Appendix p. 38, 1979.

Luhmann, N.C., Infrared and Millimeter Waves, vol. 2, K. Button editor, 1979.

MacKenzie, L.A., Masher, C.H., Dalman, G.C., 5e Congrès Tubes pour Hyperfrequence, Paris, 1964.

Phillips, T.G. and Jefferts, K.B., Rev. Sci. Instrum. 44, 1009, 1973.

Phillips, T.G., Woody, D.P., Dolan, G.J., Miller, R.E., Linke, R.A., IEEE Trans Mag-17, 684, 1981.

Poorter, thesis Univ. of Groningen, 1978.

Richards, P.L., Shen, T.-M., IEEE Trans on Electron Devices, ED27, 1909, 1980.

Rudner, S., Feldman, M.J., Kollberg, E.J., Claeson, T., J. Appl. Phys. to be published, 1981.

Schultz, G.V., 5th Int. Conf. on IR and MM Waves Würzburg, 1980.

Tuchendler, J., Grynberg, M., Couder, Y., Thomé, H., and Le Toullec, R., Phys. Rev. B, 8, 3884, 1973.

van Vliet, A.H.F., to be published in International Journal of Infrared and Millimeter Waves, 1982.

van Vliet, A.H.F., de Graauw, Th., Lidholm, S., v.d. Stadt, H., 4th Int. Conf. on IR and Submm Waves, Miami, Appendix p. 40, 1979.

Waters, J.W., Gustincic, J.J., Kakar, R.K., Roscoc, H.K., Swanson, P.N., Phillips, T.G., de Graauw, Th., Kerr, A.R., Mattauch, R.J., J. of Geophys. Research 84, 7034, 1979.

OBSERVATIONAL ASPECTS OF THE MILLIMETRE-WAVE COSMIC
BACKGROUND RADIATION

D.H. Martin
Physics Department, Queen Mary College, Mile End Road,
London E1 4NS.

1. INTRODUCTION

The millimetre-wave cosmic background radiation (CBR) offers
remarkable opportunities for investigating the early universe and a
conference on millimetre-wave astronomy might appear incomplete without
reference to it. I have been asked to review the present observational
situation and to describe what steps are being taken to begin a new wave
of more precise measurements of the spectrum of the CBR shortwards of
λ = 2 mm. [The talk presented at the Conference included a critical
review of past experiments but similar reviews have now been published -
Weiss (1980), Martin (1982a) - and for this reason this paper is
restricted to the other part of the Conference talk, i.e. a description
of the system being built at Queen Mary College as an illustration of
the kinds of problem that have to be solved if measurements are to be
made shortwards of λ = 2 mm with absolute precisions of 0.01 K or better.]

The most precise and thoroughly analysed measurement to date
is that of Woody and Richards (1981). The corrected data show significant
deviations from Planckian form; the best-fit Planckian curve corresponds
to a temperature 2.96 K but that Planck curve falls below the 1σ uncertainty of $^{+4}_{-7}$% at the peak just shortwards of λ = 2 mm, and above that at
about 1 mm. Moreover, the longer-wave microwave measurements indicate
a lower temperature, close to 2.70 K. If there were a 38% scaling error
in the calibration of the measurement of flux, the data would lie close
to the Planckian curve for 2.70 K; however, the experimenters can, after
very careful analysis of their data, find no source of an error of that
magnitude. Nevertheless they conclude that "there are serious limitations
to the statistical analysis where systematic errors are likely. It is
clear that another generation of measurements with better accuracy is
required before any deviation from a Planck spectrum can be firmly
established". Two aspects of CBR measurements in particular need to be

improved; the first is the absolute calibration procedure and the second
is the reduction of the contamination of the background by atmospheric
emission shortwards of 1.25 mm. (See Weiss, 1980 and Martin, 1982a, for
further discussion).

2. A NEW RADIOMETER FOR CBR MEASUREMENTS AT λ < 2 mm

The radiometer system described here is based on two main new
features. First, in order to be more sure that the signal detected is
not significantly contaminated by thermal emission from warmer parts of
the cryogenic enclosure, a single-mode optical system has been devised.
The detected beam is defined by a back-to-back pair of horns placed at a
quasi-focus within the system. This component has been designed so that,
over a bandwidth 1:1.25 the narrow throat at its centre passes only one
wave-guide mode, and so that the horns transform this mode into free-space
propagating beams that have spherical wave-fronts with a Gaussian
amplitude distribution in the transverse planes. The diffraction-
controlled propagation of such a beam through an optical system can be
reliably calculated, and the strong tapering of the beam suppresses
diffraction at the edges of the apertures through which the beam passes
on its way out of the cryostat.

Secondly, we have devised an absolute calibration method
based on null principles, which removes most of the inaccuracies
associated with non-ideal optical components and which also eliminates
the need for pre- or post-flight calibration of the system's responsivity.
There is a single configurational measurement mode, i.e. there is no stage
at which calibration sources have to be moved into and out of the beam.
To achieve this a temperature-active black-body reference cavity is
required; no changes in optical configuration are made, only changes in
the temperature of the reference source. Data are taken for a series of
temperatures and the null, at which the signal and reference cavities are
equal in temperature, is found by post-flight interpolation of the flight
data, for each spectral frequency. The null is insensitive to any optical
inequivalence of signal and reference channels. The (Dicke) switching
from signal channel to reference channel is achieved interferometrically
with polarisation coding to suppress any response to thermal emission
signals that do not pass right through the optical system (Martin, 1982b).

In the first flight of the system, in late 1982 or early 1983,
we shall use Ge bolometers, with a ^{3}He cooler to take the temperature to

0.3 K (Torre and Chanin, 1980); the horn-pair will be designed to cover
the wavelength range 1.8 - 2.0 mm where there is no known atmospheric
emission of significant magnitude.

In subsequent flights there will be modifications to the
radiometer that will enable us to go to wavelengths shortwards of 1.25 mm
where the atmosphere has a rich emission spectrum of closely spaced
molecular lines. To do this we plan to replace the Ge bolometers in the
radiometer with submillimetre-wave heterodyne receivers each of which
responds over an extremely narrow frequency band which will be selected
to lie between atmospheric lines. At $\sim$3.4 mbar (40 km altitude) line-
widths are of the order of 5×10^{-4} cm^{-1} and even the very closely-spaced
ozone lines are separated by more than ten line-widths. The wavelengths
at which we plan to make measurements in the first instance are 1.06 and
0.85 mm, but there are many other suitable windows between atmospheric
lines in which this approach should be successful.

The main features of the radiometer are shown in figure 1.
The system is enclosed in the helium cryostat Q, the cover R being closed
during laboratory testing and during ascent of the balloon, and open
during measurements.

H is the back-to-back pair of horns. This, together with the
off-axis ellipsoidal mirror A, forms (it is simpler to describe the
system for time-reversed propagation) a Gaussian beam which propagates
through a beam-waist centred on the aperture a, and on through aperture b,
to give a 1.6° beam on the sky. The strong taper of the beam gives only
very weak diffractive scattering at the apertures; such diffracted rays
as are formed at a are not absorbed by the warmer parts of the cryostat
but re-directed to the cold background by the shaped reflecting cone W.

The signal emerging from the other horn of H is collimated
and directed through a polarising two-beam interferometer in the
enclosure D; details of this are given below. The output beam passes into
the ^{3}He-cooler at F, which houses two detectors. The moveable reflector
in the interferometer is driven via a rigid shaft by a torque motor at t.

The radiometer is filled with He gas and surrounded by liquid
helium which, at float altitude pressure, is at 1.6 K. It has an inner
enclosure D, separated from the outer, B, by nylon sheet. It is wound
with an electrical heater and its temperature can be set, and accurately
controlled ($<\pm$0.01 K), to any value between 2.0 and 3.5 K. An internal

black cavity at C, and the back-to-back horn-pair H, are parts of this
temperature-controlled unit.

        If the temperature of this unit should be equal to the
temperature of the background at a particular spectral frequency, the
signal beam at that frequency which passes through H will be identical
to the additional thermal emission that would result if the aperture of
H were closed off (regardless of the magnitudes of the reflectance and
absorbance of H). Consequently, the interferometer is then effectively
in a black-body cavity, and movement of its mirror can give no modulation
of the detector output at the switching frequency that corresponds to
that spectral frequency. This is the basis of the null method of
calibration. This argument is not complete because it takes no account
of the fact that the detectors are not at the cavity temperature and there
could be significant reflection in the radiometer of the detectors'
emission signal, back into the detector. However, if <u>both</u> output ports
are closed with the same temperature (0.3 K) there will be no net
interferometric modulation of the reflected signals. Thus the null
detection is essentially a matter of the symmetry of the four-port optical
system. Since the black-body peak for 0.3 K is at a longer wavelength
than the wavelengths under study here, the temperatures in the output
ports need not be precisely equal.

        The null method means that calibration is part of the in-
flight measurement routine. Our approach permits, however, a pre-flight
proving of the calibration system. To do this, the cone W is replaced by
a large black-body cavity attached to a liquid-helium reservoir. Its
temperature will be varied by changing the pressure over the liquid
helium and the radiometer should then read directly the temperature of
this cavity.

3. THE ANTENNA SYSTEM

        The received beam is defined by the horn-pair at H in
figure 1 which serves as a mode-filter. For the measurements in the
wavelength range 1.8 - 2.0 mm, the horn-pair comprises two back-to-back
corrugated horns separated by a short length of corrugated wave-guide.
Corrugated guides and horns have been investigated theoretically (see
the reprint collection edited by Love, 1976) and have previously been
used in communications systems in the 10-20 GHz range. The dimensions
of the guide are such that, over this frequency band, only the HE_{11} mode
will propagate, and that with low loss. The dimensions of the horn

(length, aperture diameter, fin depth, width and spacing) are such that there should be very little reflection or mode-conversion in the horns and the HE_{11} wave-guide mode should be transformed into a plane-polarised free-space beam which is axially symmetric and has a distribution in the transverse plane that is very close to Gaussian in form, down to the -20 dB level, with very weak side-lobes outside this Gaussian region. The phase-fronts of the beam should be close to spherical not only in the far-field but also in the middle-field where mirrors pick-up the beams, and the centre of curvature should change position by no more than a millimetre or two over the measurement bandwidth. The advantage of such beams is that it is possible to analyse their diffraction-controlled free-space propagation (see below).

We fabricate such horn-pairs by electro-forming and we are currently assessing their performance in a millimetre-wave antenna test facility. The measured antenna patterns can be in excellent accord with theoretical predictions. To measure the Gaussian-beam transmission and reflection properties of the horn-pairs we are using procedures that are free-space equivalents of the wave-guide balanced-circuit methods used at lower frequencies.

The diverging beam launched (received) by the horn-pair is converted into a converging beam by the off-axis ellipsoidal reflector, shown at A in figure 1. This has the dimensions that would, in geometrical optics, produce a focus near the aperture a and a diverging beam passing through the aperture b, to give a 1.6^{o} beam on the sky. For $\lambda \sim 2$ mm, however, the propagation will be diffraction controlled; a Gaussian beam remains Gaussian as it propagates, albeit with a varying beam-width and with a shifting phase-centre for the wave-front (Martin and Lesurf, 1978; Lesurf, 1981). The profile and diameter of the reflector at A and the diameters of the apertures a and b, have been

	a Diameter 72 mm			b Diameter 254 mm		
Wavelength (mm)	1.9	1.5	2.3	1.9	1.5	2.3
Beam-width (mm)	28.6	28.2	28.6	78.2	71.8	85.0
Aperture/Beamdwidth	2.52	2.55	2.52	3.24	3.54	3.00

determined in the light of such analyses so as to ensure that the beam amplitudes at the edges of the apertures are strongly tapered in order to suppress edge-diffraction there. The reflector R truncates the beam outside its -20 dB points (the horn is used as nearly on-axis as is consistent with this requirement, in order to minimise the off-axis distortion of the Gaussian beam). The table compares the calculated Gaussian beam-widths ($^{1}/e$, amplitude) at the apertures a and b with the aperture diameters. The ratio is greater than 2.5 in each case, corresponding to a tapering at the aperture edges in excess of -54 dB.

The GTD method of calculating edge-diffraction effects (see the reprint collection edited by Hansen, 1980) has been used to calculate contributions to the system's antenna temperature attributable to the residual diffraction at the aperture edges. First there is the near-field diffraction into the several, successively hotter, regions within the cryostat (T and W in figure 1). Note that the cone W is highly reflecting (metallised fibre-glass) and is shaped so that any ray entering it through a is reflected out of the cryostat within 16° of the axis, thereby avoiding the gondola, balloon and Earth.

Secondly there is the far-field large-angle antenna pattern external to the cryostat. The diameters of the apertures at A, a and b are such that, to emerge from the cryostat at an angle larger than 16° to the axis, a ray would have to suffer successive diffraction at all three apertures. The calculated rejection ratio, relative to the on-axis beam, exceeds 10^{9} at angles greater than 20°, of which 10^{2} comes from the tapering of the beam at the edges of R.

The more sensitive aspects of the antenna design, as revealed by the calculations, are currently being checked using the millimetre-wave test facility.

The gondola will be carried on a 600 m nylon cable below the balloon. The balloon then subtends a half-angle of 5° at the gondola and lies without 20° of the radiometer axis; the Earth lies without 65° of the axis.

The main beam itself intercepts a very thin polymer film covering the aperture at b the purpose of which is to prevent condensation of air in the radiometer. At $\lambda \approx 2$ mm the reflectance and emissivity of such a film are extremely small. The film is belled out by the small excess pressure in the cryostat and is shaped to have a centre-of-curvature within the aperture a, thereby reflecting back the cold signal from a.

Our overall calculations indicate that the total contribution
to the antenna temperature from all objects other than the cosmic background should be little more than 0.01 K. Most of this originates
internally and should be detected as an off-set when proving the
calibration in the laboratory, as described earlier.

The discussion above relates to the measurements to be made
at $\lambda \approx 2$ mm. For subsequent measurements at shorter wavelengths the same
beam-forming system will be used except for replacement of the horn-pair.
Diffraction should be a lesser problem at shorter wavelengths and adequate
beam-control should be obtained with a smooth-walled horn-pair with
flared apertures (Mather, 1981); it might be difficult to fabricate a
low-loss, low-reflection corrugated horn-pair for the shorter wavelengths.

4.   INTERFEROMETRIC DICKE SWITCH

The unit enclosed in the box D of figure 1 is a polarising
two-beam interferometer used as a Dicke switch (Martin, 1982). The lay-
out is shown in figure 2. The horn-pair H passes through the base plate
and the beam emerging from it illuminates the off-axis ellipsoidal
collimator J, passes through the wire-grid polariser K, and is then split
into equal, orthogonally polarised, beams at the wire-grid beam-divider
G. One of the two beams returns to G via the roof-edge reflector M, and
the other via the corresponding reflector N; they are re-combined there
to pass out of the interferometer via the reflectors X and Y. Reflector
M moves along its optical axis on a ball movement; it is driven via a
rigid tubular shaft from a second ball movement with integral torque
motor, mounted at the top of the cryostat. The grids K and G are
produced by winding 10 μm gold-coated tungsten wire onto invar frames
under controlled light tension.

As the moving reflector is displaced at constant speed, the
detector in the output registers, alternately, firstly the signal beam
entering by transmission through the polariser K and, second, the
orthogonally polarised signal reflected from K due to thermal emission
by the cavity B. The walls of B are absorbing and it serves as the
reference source for calibration (see below). The switching is produced
interferometrically; since it consequently has a frequency inversely
proportional to the wavelength of the signal, the time-dependent
detector output can be Fourier analysed to separate spectral components
of the input beam.

The reflector J places beam-waists near N and M in the interferometer; it does not have to be as wide as the ellipsoidal reflector
in the antenna system (nor does the entrance to B) because the whole of
the system shown in figure 2 is at a uniform temperature (see below) and,
as a result, weak diffractive spillage into the beams from the box
enclosing this unit will not seriously compromise the calibration versus
B. Note that only beams passing through K will be interferometrically
modulated. The weak reflection of the cold-signal from the detector port
by the horn-pair at H can be balanced if necessary by setting a wire-grid
polariser at the entrance to the cavity B at an appropriate offset from
parallelism with the grid K, as determined in bench tests by nulling the
signal in one output port due to a bright source mounted in the second
output port.

The speed of movement of M must be such as to give switching
frequencies in the range appropriate to the response-times of the
detectors, the sampling intervals must be small enough to avoid aliasing,
and the maximum mirror displacement must be sufficient to give the
required spectral resolution which, in turn, is limited by the need to
obtain an adequate signal-to-noise ratio in the time available for taking
measurements. These considerations lead us to the following values for
the operating parameters, corresponding to a spectral resolving power of
25 at $\lambda \approx 2$ mm.

Reflector speed: 20 mm s^{-1}
Maximum movement: ±25 mm
Sampling rate: 400 Hz
Corner frequency for 4-pole filter: 80 Hz

5.   THERMAL CONTROL FOR CALIBRATION

The null method of calibration requires that it be possible to
set the temperature of the reference cavity B, and the horn-pair H, to a
succession of values in the range 2.0 - 3.5 K and to maintain that temperature uniform, and constant over the time of a measurement, to within the
required precision of the CBR measurement. The antenna system should be
good to 0.01 K and the thermal design is based on a similar precision,
over measurement times of tens of minutes. It proves to be practicable
to exercise such control of temperature for the whole of the interferometer, not simply for the cavity B and the horn-pair H, and that
allows a more compact optical design. The temperature-controlled unit,
shown enclosed in a double-walled box B-D in figure 1, thus includes all

the components shown in figure 2. The box contains gaseous helium but
the outer wall is sealed with indium-wire gaskets to exclude liquid helium.
The inner is supported from the outer wall by nylon brushes, and the space
between the walls is filled out with nylon sheet to suppress convective
motion of the He gas. The unit incorporates an electrical heater. The
thermal time constant for the unit is about 2 s which is convenient for
controlling the temperature over measurement times of several minutes;
the power supplied to maintain the maximum temperature of 3.5 K gives an
acceptably low contribution to the liquid helium boil-off rate ($<\frac{1}{2}\ell s^{-1}$).

        Most of the components in the unit, and the enclosing boxes,
are made of high-conductivity copper to promote uniformity of temperature
in the steady-state, and rapid attainment of the steady-state after an
injection of heat for temperature control. The decay of a deviation from
a steady-state distribution of temperature will be characterised by a
time-constant equal to $\frac{\ell^2}{16K'}$ where K' is the thermal diffusivity of the
copper and ℓ is a measure of the spatial scale of the deviation. For
high-conductivity copper, and for $\ell \sim 0.1$ m, the time-constant is no more
than a few milliseconds which is acceptably fast. In the steady-state
there would be significant non-uniformity of temperature if the location
of the heat input were remote from that of the heat leak and both the
input and the leak are roughly uniformly distributed over the outer face
of the inner wall for this reason.

        The cavity B must be black and its emission temperature well-
defined. The surfaces are specular and at least five reflections would be
suffered before an incident beam re-emerged. The temperature difference
between cavity and detector is 3.5 - 0.3 K, and the blackness must therefore be better than 99% if the calibration uncertainty is to be less than
0.01 K without correction. A reflectance of 30% at each surface suffices
and is provided by a 1 mm layer of Eccosorb CR110 (Emerson & Cuming Ltd.)
bonded to a copper backing-plate. Since it is the temperature of the
metal backing plate that will be measured, it is necessary that the
temperature throughout the absorbing layer be well within 0.01 K of that
of the backing plate. This matter requires consideration because the
absorbing must have a thickness approaching $\lambda/4$, or more, if it is to
absorb significantly in the presence of the metal backing. The thermal
load on a surface in B is extremely small, however, and calculation
shows that a temperature difference of <0.01 K should be readily achieved
(nevertheless, we are making measurements of the thermal conductance

across bonded sandwiches of copper and Eccosorb CR110 to confirm this).
This conclusion would be disturbed if thermal emission from the atmosphere
were allowed to pass through the antenna system to become a thermal load
on the cavity B. This emission is mainly at wavelengths around 13 µm.
Most of this will be absorbed at the walls of tube T or in a thin IR-
absorbing layer on the reflector at A. Moreover, the wave-guide section
in the horn-pair will serve as a small aperture and the wire-grid K will
strongly scatter the residual power into the large interferometer cavity.
Finally, the condensing mirror at the aperture of B will also carry an
IR-absorbing coating. The longer-wave infrared emission from the atmosphere is much weaker, and it will be partially transmitted by K, and
partially reflected into the subsidiary absorbing cavity shown in figure 2,
and will not constitute a significant thermal load on the cavity B.

Thermometry is of importance both for the sensors of the
control system and for the absolute measurement of the cavity temperature.
Carbon-glass thermometers are used for both purposes and in addition a
^{4}He vapour-pressure thermometer is incorporated to provide an absolute
measure of the temperature directly by reference to the practical standard
of temperature. Such a thermometer, with a commercially available
pressure-transducer, is well suited to measurements in the range 2.0 -
3.5 K, to a precision of a few mK.

A full-scale thermal model of the calibration system has been
built and has confirmed the required thermal behaviour under control.

6.   THE ^{3}He COOLER AND DETECTORS

Our collaborators at Service d'Aeronomie, CNRS, Paris, have
provided a ^{3}He-cooler (Torre and Chanin, 1980) for the detectors. This
is a compact unit, with the ^{3}He in a closed high-pressure vessel,
comprising a cryopump bulb linked by a tube to a smaller bulb into which
the ^{3}He condenses, mounted inside a vacuum jacket. Each bulb is
connected through a ^{4}He heat-pipe (heat switch) to the 1.6 K ^{4}He bath in
which the cooler is immersed. The cooler is initiated, by a 30 minute
controlled sequence of inputs to electrical heaters on the cryopump and
heat-switches, and then operates for >8 hours at a temperature of 0.3 K.

Germanium bolometers cooled to 0.3 K have noise-equivalent-
powers in the range 10^{-15}-10^{-14} WHz$^{\frac{1}{2}}$ and, when matched to a Gaussian beam,
this gives a temperature resolution better than 0.01 K in a 15 minute
integration time for a spectral resolving power of 25 at $\lambda \approx 2$ mm. The
Gaussian beam emerging ~rom the interferometer unit is condensed by a

silica lens to a beam-waist at the detector, which is mounted on the ^{3}He
bulb. This lens is bloomed with a PTFE film and incorporates a band-pass
filter; it forms the vacuum window of the ^{3}He-cooler. At a beam-waist
the phase-front is plane and it is therefore possible to use a reflector,
at a distance $\lambda/4$ behind the detector, to optimally couple the signal
power into the absorbing film of the bolometer. A wire-grid polariser
(the final component of the polarising interferometer) is mounted inside
the cooler in front of the detector. In fact, a second bolometric
detector is used to record the complementary signal reflected from the
polariser, which is the second output port of the interferometer.

For measurements of the CBR at wavelengths $\lesssim 1$ mm we plan,
as explained earlier, to use heterodyne receivers in place of the
bolometers. A number of new problems have to be solved. Solid-state
local oscillators, and quasi-optical four-port couplers to give balanced-
mixer configurations, are required; the development of these is raising
a number of interesting technical problems.

ACKNOWLEDGMENTS

The following colleagues have been engaged in developing
the radiometer described here under a grant from the UK Science and
Engineering Council: E.F. Puplett, B. Rose, P. Carnie, R. Wylde.
Collaborators at the Polytechnic of Central London (P. Morse and
D. Alston) are providing the control and data system for the
measurements and those at the CNRS Service d'Aeronomie, Paris (G. Chanin
and J.P. Torre) the ^{3}He-cooler and bolometers.

REFERENCES

Hansen, R.C., 1980, Ed: Geometric Theory of Diffraction, IEEE Press
          Selected Reprint Series.
Lesurf, J.C.G., 1981, Infrared Physics, 21, 383.
Love, A.W., 1976, Ed: Electromagnetic Horn Antennas, IEEE Press
          Selected Reprint Series.
Martin, D.H., 1982a, in "Progress in Cosmology" pp.119-143,
          Ed. A.W. Wolfendale, Reidel.
Martin, D.H., 1982b, Ch.4 of Millimetre and Submillimetre Waves,
          Ed. K.J. Button, Academic Press.
Martin, D.H. & Lesurf, J.C.G., 1978, Infrared Physics, 18, 405.
Mather, J.C., 1981, IEEE Trans. AP29, 967.
Torre, J.P. & Chanin, G., 1980, Proc.ICEC, 8, 112.
Weiss, R., 1980, Ann.Rev.Astron.Astrophys. 18, 489.
Woody, D.P. & Richards, P.L., 1981, Astrophys.J. 248, 18.

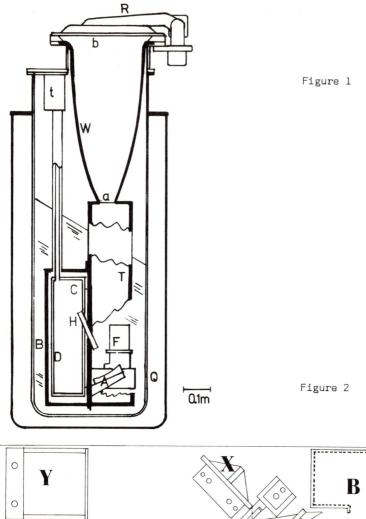

Figure 1

Figure 2

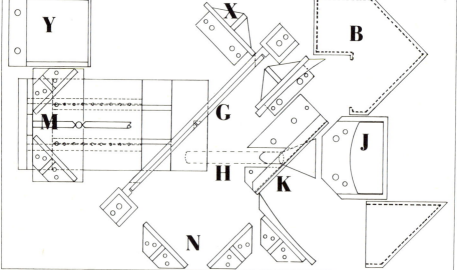

INDEX